ROBERT PLOMIN

Blueprint

How DNA Makes Us Who We Are

The MIT Press
Cambridge, Massachusetts
London, England

This book was set in Sabon LT Std by Jouve (UK), Milton Keynes. Printed and bound in the United States of America.

Library of Congress Cataloging-in-Publication Data
Names: Plomin, Robert, 1948- author.
Title: Blueprint : how DNA makes us who we are / Robert Plomin.
Description: Cambridge, MA : The MIT Press, [2018] | Includes bibliographical references and index.
Identifiers: LCCN 2018032148 | ISBN 9780262039161 (hbk. : alk. paper)
Subjects: | MESH: Genetics, Behavioral | DNA
Classification: LCC QH438.7 | NLM QU 450 | DDC 572.8/6--dc23 LC record available at https://lccn.loc.gov/2018032148

10 9 8 7 6 5 4 3 2 1

Contents

Prologue

What would you think if you heard about a new fortune-telling device that is touted to predict psychological traits like depression, schizophrenia and school achievement? What's more, it can tell your fortune from the moment of your birth, it is completely reliable and unbiased – and it costs only £100.

This might sound like yet another pop-psychology claim about gimmicks that will change your life, but this one is in fact based on the best science of our times. The fortune teller is DNA. The ability to use DNA to understand who we are, and predict who we will become, has emerged only in the last three years, thanks to the rise of personal genomics. We will see how the DNA revolution has made DNA personal by giving us the power to predict our psychological strengths and weaknesses from birth. This is a game-changer that has far-reaching implications for psychology, for society and for each and every one of us.

This DNA fortune teller is the culmination of a century of genetic research investigating what makes us who we are. When psychology emerged as a science in the early twentieth century, it focused on the environmental causes of behaviour. Environmentalism – the view that we are what we learn – dominated psychology for decades. From Freud onwards, the family environment, or *nurture*, was assumed to be the key factor in determining who we are. In the 1960s geneticists began to challenge this view. Psychological traits from mental illness to mental abilities clearly run in families, but there was a gradual recognition that family resemblance could be due to *nature*, or genetics, rather than nurture alone, because children are 50 per cent similar genetically to their parents.

Since the 1960s scientists conducting long-term studies on special relatives like twins and adoptees have built a mountain of evidence showing that genetics contributes importantly to psychological differences between us. The genetic contribution is not just statistically significant, it is massive. Genetics is the most important factor shaping who we are. It explains more of the psychological differences between us than everything else put together. For example, the most important environmental factors, such as our families and schools, account for less than 5 per cent of the differences between us in our mental health or how well we did at school – once we control for the impact of genetics. Genetics accounts for 50 per cent of psychological differences, not just for mental health and school achievement, but for all psychological traits, from personality to mental abilities. I am not aware of a single psychological trait that shows no genetic influence.

The word 'genetic' can mean several things, but in this book it refers to differences in DNA sequence, the 3 billion steps in the spiral staircase of DNA that we inherit from our parents at the moment of conception. It is mind-boggling to think about the long reach of these inherited differences that formed the single cell with which we began life. They affect our behaviour as adults, when that single cell with which our lives began has become trillions of cells. They survive the long and convoluted developmental pathways between genes and behaviour, pathways that meander through gene expression, proteins and the brain. The power of genetic research comes from its ability to detect the effect of these inherited DNA differences on psychological traits without knowing anything about the intervening processes.

Understanding the importance of genetic influence is just the beginning of the story of how DNA makes us who we are. By studying genetically informative cases like twins and adoptees, behavioural geneticists discovered some of the biggest findings in psychology because, for the first time, nature and nurture could be disentangled. The implications of these findings are transformative for psychology and society and for the way you think about what makes you who you are.

For example, one remarkable discovery is that even most measures of the environment that are used in psychology – such as the quality of parenting, social support and life events – show significant genetic impact. How is this possible when environments have no DNA themselves? As we shall see, genetic influence slips in because these are not pure measures of the environment 'out there' independent of us and our behaviour. We select, modify and even create our experiences in part on the basis of our genetic propensities. This means that correlations between such so-called 'environmental' measures and psychological traits cannot be assumed to be caused by the environment itself. In fact, genetics is responsible for half of these correlations. For example, what appears to be the environmental effect of parenting on children's psychological development actually involves parents responding to their children's genetic differences.

A second crucial discovery at the intersection of nature and nurture is the unexpected way in which the environment makes us who we are. Genetic research provides the best evidence we have for the importance of the environment because genetics accounts for only half of the psychological differences between us. For most of the twentieth century environmental factors were called *nurture* because the family was thought to be crucial in determining who we become. Genetic research showed that this is absolutely not true. In fact, the environment makes siblings reared in the same family as different as siblings reared in separate families. Family resemblance is due to our DNA rather than to our shared experiences like TLC, supportive parenting or a broken home. What makes us different environmentally are random experiences, not systematic forces like families. The implications of this finding are enormous. Such experiences affect us, but their effects do not last; after these environmental bumps we bounce back to our genetic trajectory. Moreover, what look like systematic long-lasting environmental effects are often reflections of genetic effects, caused by us creating experiences that match our genetic propensities.

As I will demonstrate in this book, the DNA differences inherited from our parents at the moment of conception are the consistent, lifelong source of psychological individuality, the blueprint that makes us who we are. A blueprint is a plan. It is obviously not the same as the finished three-dimensional structure – we don't look like a double helix. DNA isn't all that matters but it matters more than everything else put together in terms of the stable psychological traits that make us who we are.

These findings call for a radical rethink about parenting, education and the events that shape our lives. The first part of *Blueprint* concludes with a new view of what makes us who we are that has sweeping and no doubt controversial implications for all of us. It also provides a novel perspective on equal opportunity, social mobility and the structure of society.

These big findings were based on twin and adoption studies that indirectly assessed genetic impact. Twenty years ago the DNA revolution began with the sequencing of the human genome, which identified each of the 3 billion steps in the double helix of DNA. We are the same as every other human being for more than 99 per cent of these 3 billion DNA steps, which is the blueprint for human nature. The less than 1 per cent of these DNA steps that differ between us is what makes us who we are as individuals – our mental illnesses, our personalities and our mental abilities. These inherited DNA differences are the blueprint for our individuality, which is the focus of the second part of *Blueprint*.

Recently, it has become possible to directly assess each of the millions of inherited DNA differences between us and to find out which of these are responsible for the ubiquitous genetic influence on psychological traits. One of the extraordinary discoveries was that we are not just looking for a few DNA differences with big effects but rather thousands of small differences whose weak effects can be aggregated to create powerful predictors of psychological traits. The best predictors we have so far are for schizophrenia and school achievement, but other DNA predictors of psychological traits are being reported every month.

These are unique in psychology because they do not change during our lives. This means that they can foretell our futures from birth. For example, in the case of mental illness, we no longer need to wait until people show brain or behavioural signs of the illness and then rely on asking them about their symptoms. With DNA predictors we can predict mental illness from birth, long before any brain or behavioural markers can be detected. In this way, DNA predictors open the door to prediction and, eventually, prevention of these problems before they create collateral damage that is difficult to repair. These DNA predictors are also unique in genetics because for the first time we can go beyond predicting the average risk for different members of a family to predict risk separately for each member of the family. This is important because family members differ a lot genetically – you are 50 per cent similar genetically to your parents and siblings, but this means that you are also 50 per cent different.

These new DNA developments are described in the second part of *Blueprint*, which concludes by showing how this new era of DNA predictors will transform psychology and society – and how we understand ourselves. The applications and implications of DNA predictors will be controversial. Although we will examine some of these concerns, I admit I am unabashedly a cheerleader for these changes. At any rate, the genome genie is out of the bottle and cannot be stuffed back in again.

Blueprint focuses on psychology for two reasons. First, psychology is the essence of who we are, our individuality. Most of the same conclusions apply to other sciences such as biology and medicine, but the implications of the DNA revolution are more personal for psychology.

A second reason is that I am a psychologist who has for forty-five years been at the centre of genetic research on mental health and illness, personality and mental abilities and disabilities. One of the best things in life is to find something that you love to do, and I fell in love with genetics when I was a graduate student in psychology at the University of Texas at Austin in the early 1970s. It was thrilling to be part of the beginning of the modern era of genetic research in psychology. Everywhere we looked we found evidence for the importance of genetics, which was amazing, given that genetics had been ignored in psychology until then. I feel lucky to have been in the right place at the right time to help bring the insights of genetics to the study of psychology.

I have been waiting thirty years to write *Blueprint*. My excuse for not doing it sooner is that more research was needed to document the importance of genetics, and I was busy doing that research. However, in hindsight, I have to admit to another reason: cowardice. It might seem unbelievable today, but thirty years ago it was dangerous professionally to study the genetic origins of differences in people's behaviour and to write about it in scientific journals. It could also be dangerous personally to stick your head up above the parapets of academia to talk about these issues in public. Now, the shift in the zeitgeist has made it much easier to write this book. A huge bonus for waiting is that the story is much more exciting and urgent now because the DNA revolution has advanced in ways no one anticipated thirty years ago. Now, for the first time, DNA by itself can be used to make powerful predictions of who we are and who we will become.

Blueprint interweaves my own story and my DNA in order to personalize the research and to share the experience of doing science. I hope to give you an insider's view of the exciting synergies that came from combining genetics and psychology, culminating with the DNA revolution. Although this book expresses my subjective view of how DNA makes us who we are, I have tried my best to present the research honestly and without hype. However, as I move further from the data to explore the implications of these findings, some issues will be controversial. My goal is to tell the truth as I see it, without pulling punches for the sake of perceived political correctness.

My focus on the importance of inherited DNA differences is likely to attract criticism for resurrecting the nature versus nurture debate long after its widely reported demise. Throughout my career I have emphasized nature *and* nurture, not nature *versus* nurture, by which I mean that both genes and environment contribute to the psychological differences between people. Recognition that both genes and environment are important fosters research at the interplay between nature and nurture, a very productive area of study.

However, the problem with the mantra 'nature and nurture' is that it runs the risk of sliding back into the mistaken view that the effects of genes and environment cannot be disentangled. No one has trouble accepting that the environment we experience contributes to who we are, but few people realize how important DNA differences are. My reason for focusing on DNA as the blueprint for making us who we are is that we now know that DNA differences are the major systematic source of psychological differences between us. Environmental effects are important but what we have learned in recent years is that they are mostly random – unsystematic and unstable – which means that we cannot do much about them.

I hope *Blueprint* launches a conversation about these issues. A good conversation requires DNA literacy, which this book attempts to provide, especially in relation to complex psychological traits. This requires some knowledge about DNA, the statistics of individual differences, and the technological advances that have led to the DNA revolution. I have attempted to explain these complicated ideas as simply as possible. A 'Notes' section at the end of the book provides references and additional explanation for these and other topics. Because the issues tackled in *Blueprint* are more than complicated enough, I have resisted digressions into research on topics that, although fascinating, are not essential to understanding inherited DNA differences as they relate to psychological traits. Some of these tangential topics that I have reluctantly let go include evolution, epigenetics and gene editing.

I hope this book conveys the excitement I feel about this historic moment in psychology. The message from earlier research has begun to sink in, that DNA is the major systematic force, the blueprint, that makes us who we are. The implications for our lives – for parenting, education and society – are enormous. However, this only sets the stage for what will be the main event: the ability to predict our psychological problems and promise from DNA. This is the turning point when DNA changes psychology – scientifically and clinically – and the impact of psychology on our lives. Our future is DNA.

PART ONE

Why DNA matters

I

Disentangling nature and nurture

We are all similar in many ways. With few exceptions, we stand on two feet, we have eyes in the front of our heads that allow us to see in three dimensions and, most amazingly, we learn to speak. But we are also obviously different – physically, physiologically and psychologically. *Blueprint* is about what makes us different psychologically.

Psychologists study hundreds of traits, which is their collective label for differences between us that are consistent across time and across situations. These traits include dimensions of personality, such as emotionality and energy level, and traits that are traditionally assessed as either-or disorders, for instance depression and schizophrenia. They also include cognitive traits such as general learning ability, often called intelligence, and specific mental abilities such as vocabulary and memory, as well as disabilities in these traits.

For most of the twentieth century it was assumed that psychological traits were caused by environmental factors. These environmental factors were called *nurture* because, from Freud onwards, their origins were thought to lie in the family environment. Because these traits run in families, it was reasonable to assume that the family environment is responsible for these traits.

But genetics also runs in families. Fifty years before we knew about DNA we knew that first-degree relatives – parents and their children, brothers and sisters – are 50 per cent similar genetically. So the reason why psychological traits run in families could be nature (genetics) as well as nurture (environment). However, it is more difficult to credit nature because DNA is invisible and silent but you can see, hear and feel the nurture of family life, for good and for bad.

So, what is the relative importance of nature and nurture for psychological traits? First, take a minute to note your opinions about nature (genetics) and nurture (environment). By rating the following traits now, you can then compare your ratings to those of other people and to the results of genetic research. Although this book is about psychological traits, it is useful to begin by contrasting psychological traits with a few physical traits (eye colour, height) and medical traits (breast cancer, stomach ulcers).

For the following fourteen traits, rate how much you think genetic factors are important in making people different – in other words, how heritable do you think they are? If you think that a trait shows no genetic influence, rate it as 0 per cent. If you think that a trait is entirely due to genetic influence, rate it as 100 per cent. For some of the traits, you might not have any idea about how much DNA matters, but make a guess.

On page 6 you can compare your ratings to those from a 2017

Table 1 To what extent (from 0% to 100%) do you think the following traits are heritable?

Eye colour	_____
Height	_____
Weight	_____
Breast cancer	_____
Stomach ulcers	_____
Schizophrenia	_____
Autism	_____
Reading disability	_____
School achievement	_____
Verbal ability	_____
Remembering faces	_____
Spatial ability (e.g., navigation)	_____
General intelligence (e.g., reasoning)	_____
Personality	_____

survey of 5,000 young adults in the UK. The last column shows estimates based on decades of genetic research which indicate that inherited DNA differences account for about 50 per cent of our psychological differences. In other words, inherited DNA differences are the main reason why we are who we are. The next chapter explores how we know this to be true, and the rest of the first part of *Blueprint* investigates what it means for psychology and society.

These fourteen traits were not selected because they are especially heritable. Substantial genetic influence has been found not only for schizophrenia and autism but for all types of psychopathology, including mood disorders, anxiety disorders, attention-deficit disorders, obsessive-compulsive personality disorder, antisocial personality disorders and drug dependence. Substantial genetic influence is also found for all aspects of personality and mental abilities and disabilities.

In fact, it is no longer interesting to show that another psychological trait is heritable, because all psychological traits are heritable. A sign of how much the situation has changed from the last century's environmentalism is that I do not know of a single psychological trait that does *not* show genetic influence.

Estimates of genetic influence are called *heritability*, which has a precise meaning in genetics. Heritability describes how much of the differences between individuals can be explained by their inherited DNA differences. The word 'differences' is key to its definition. *Blueprint* is about what makes us different psychologically.

There are many related words that create confusion around heritability. 'Innate' and 'inborn' refer to universal characteristics that are so important evolutionarily that they do not vary, at least given the range of environments in which we evolved. We all walk on two legs, we all have eyes in the front of our heads to perceive depth, and we all have basic reflexes like blinking our eyes in response to a puff of air. These characteristics are programmed by the 99 per cent of our DNA that does not differ between us. In contrast, heritability is about the 1 per cent of DNA that differs between us and contributes to our differences in behaviour. Even though innate characteristics are programmed by DNA, we can't talk about their heritability because innate characteristics do not vary between us.

Words like 'genetic' and 'inherited' – and colloquial phrases like 'in my genes' or 'in your DNA' – cover anything to do with DNA. They include the universal 99 per cent of our DNA as well as the 1 per cent that makes us different. They also include DNA mutations that are not inherited or passed on to our offspring, such as the DNA mutations in skin cells that cause skin cancer.

In science, when words have multiple meanings and connotations, it is useful to coin a new word that means only what you want it to mean. That is the reason for the six-syllable mouthful 'heritability'. It indexes the extent to which a trait like weight is heritable. The 70 per cent heritability for weight means that 70 per cent of the differences between people in their weight can be attributed to differences between them in inherited DNA sequence. The other

Table 2 How much are these traits influenced by genetics? The first column of results shows the average opinions of 5,000 young adults in the UK. The second column shows results from genetic research.

	Average ratings of 5,000 UK adults	Results of genetic research
Eye colour	77%	95%
Height	67%	80%
Weight	40%	70%
Breast cancer	53%	10%
Stomach ulcers	29%	70%
Schizophrenia	43%	50%
Autism	42%	70%
Reading disability	38%	60%
School achievement	29%	60%
Verbal ability	27%	60%
Remembering faces	31%	60%
Spatial ability (e.g., navigation)	33%	70%
General intelligence (e.g., reasoning)	41%	50%
Personality	38%	40%

30 per cent could be due to systematic environmental factors like diet and exercise, but, as we shall see, what makes us different environmentally are unsystematic, random experiences over which we have little control.

Heritability is frequently misunderstood. For example, it is not a constant like the speed of light or gravity. It is a statistic that describes a particular population at a particular time with that population's particular mix of genetic and environmental influences. A simpler way of expressing this is that it describes *what is* but does not predict *what could be*. Another population, or the same population at a different time, could have a different mix of genetic and environmental influences. Heritability will reflect these differences. For example, heritability of body weight is greater in wealthier countries such as the US than in poorer countries such as Albania and Nicaragua. Wealthier countries have greater access to fast-food outlets and high-energy snacks, and greater access to fattening food leads to higher heritability because it exposes genetic differences in people's propensities to put on the pounds.

Several other common misunderstandings about heritability stem from this confusion between *what is* and *what could be*, and from thinking about a single individual rather than individual differences in a population. (If you're interested, you can see a further discussion of this in the Notes section at the end of the book.) For now, the point of the summary of genetic research shown in Table 2 is that genetics contributes substantially to differences between people.

How did your ratings stack up against the summary of genetic research? The 'average ratings' in Table 2 show that most people accept a role for genetic influence. However, there are some large discrepancies between what most people think and what research tells us, and it is revealing to explore these discrepancies.

The biggest discrepancy is for breast cancer. On average, people think that breast cancer is mostly (53 per cent) heritable, but research shows that it is by far the least heritable of the fourteen traits (10 per cent). In other words, why do some women get breast cancer and others do not? Genetics is only 10 per cent of the answer.

One slice of the evidence makes this clear: A woman who has an identical twin with breast cancer is only at slightly greater risk of having breast cancer, even though identical twins are like clones in that they inherit the same DNA. The rate of breast cancer for women is about 10 per cent. But the rate of breast cancer for women who have an identical twin with breast cancer is only 15 per cent. Although this represents a 50 per cent increase in relative risk, in absolute terms this means that 85 per cent of the time, when one identical twin has breast cancer, the co-twin will not have breast cancer. Because identical twins are identical genetically, their discordance for breast cancer must be due to environmental differences.

We don't know what these important environmental differences are. They could be systematic factors like diet, lifestyle or illness, but they could also be due to non-inherited mutations that pop up by chance in particular cells in the breast. But the important message from this genetic research is that heritability is very low for breast cancer.

Why do people think breast cancer is so much more heritable than it is? Most people say they rated breast cancer as highly heritable because they heard about genes being found for breast cancer. It is true that a few inherited DNA differences have been found that are associated with breast cancer, but these DNA variants are very rare and have little effect on the population as a whole.

Although breast cancer is one of the least heritable traits, it is often caused by DNA differences, but these are DNA differences that are not inherited. When geneticists say a trait is heritable, they are referring to inherited DNA differences. This is in line with what people mean when they say that eye colour is highly heritable – you inherit it from your parents. This is a very narrow definition of genetic influence because it excludes many other DNA differences that are not inherited. Breast and many other cancers are triggered by DNA mutations that happen by chance in a particular somatic cell like a breast cell. We don't inherit these DNA mistakes from our parents and we don't pass them on to our children.

In contrast to this narrow but specific definition of 'genetic' as inherited DNA differences, environmental influence is defined very broadly to mean all influences that are not due to inherited DNA differences. This definition of environment is much broader than the typical environmental influences that are studied by psychologists such as family, neighbourhood, school, peer and work environments. As in the case of breast cancer, it even includes DNA differences that are not inherited. This broad definition of environment also includes prenatal influences, illnesses, and food and drink – everything and anything that is not caused by inherited DNA differences. In this sense, a better word for what geneticists mean when they refer to environment is 'non-genetic'.

The next two biggest discrepancies between what people think about heritability and what research tells us are for body weight and stomach ulcers. These discrepancies are in the opposite direction from breast cancer in that people think weight and ulcers are the least heritable physical traits but research tells us that these are among the most heritable traits. On average, people in our survey rated weight as 40 per cent heritable and ulcers as 29 per cent heritable. But genetic research shows heritability estimates of 70 per cent for both weight and ulcers.

When you ask people why they rated weight and ulcers as less heritable than the other traits, they say that weight is a matter of willpower and that ulcers are caused by stress. Willpower and stress are assumed to be driven environmentally. But these assumptions are wrong and it is important to know why.

For weight, the reason why people assume that willpower is key is that, if you stop eating, naturally, you will lose weight. Our culture often blames people who are overweight, as though they lack the self-control to stop eating. However, finding that 70 per cent of the differences between people in body weight are caused by inherited DNA differences between them does not contradict the truism that anyone can lose weight if they stop eating altogether. Anyone will also lose weight if they suddenly have no access to food or if they are fitted with gastric bands that restrict the amount of food they can eat. As we have seen, the focus of genetic research is not what *can* make a difference but rather what *does* make a difference in the population. That is, genetic research describes *what is* rather than predicting *what could be*.

What the heritability of 70 per cent for weight means is that on average the differences in weight between people that you see around you are largely due to inherited DNA differences, despite individual differences in dieting, exercise and lifestyles. Some people find it much easier to put on weight, and much harder to lose it, for genetic reasons.

Similarly, there is no evidence for the common assumption that stomach ulcers are caused by stress. Stomach ulcers are in fact often caused by bacterial infection, but this does not imply that DNA differences are unimportant. Genetics matters a lot when it comes to differences in susceptibility to infection, just as genetic influences on susceptibility to food cues can affect body weight. Genetically driven differences in susceptibility to the environment are important mechanisms by which genetic differences create differences between us biologically and psychologically.

What about psychological traits? For the last nine traits in the list, the average rating is 36 per cent, which is substantial, although considerably lower than the average research estimate of 58 per cent.

One of the biggest discrepancies between people's ratings and research results is for school achievement, which is a focus of my research. The average rating in our survey was 29 per cent, but genetic research consistently shows that performance on tests of school achievement is 60 per cent heritable on average. That is, more than half of the differences between children in how well they do at school is due to inherited DNA differences.

These average ratings mask a wide range of opinions. The widest range emerged for psychological traits. For example, the average rating for autism was 42 per cent, but 6 per cent of the sample thought autism was 100 per cent heritable and 14 per cent thought it was not at all heritable.

If you underestimated genetic influence on psychological traits, you are not alone. There is a wide range of opinion about genetic influence on psychological traits. Overall, 15 per cent of the sample rated these traits as not at all heritable.

Are some people 'environmentalists', thinking that none of these traits show genetic influence, and others 'hereditarians', believing that everything is heritable? This was not the case. People who thought one trait was highly heritable were not the same people who thought the same way about other traits.

The results of this survey were crucial in deciding how I would write this book. In the past, when psychologists and the public as a whole did not yet accept the importance of genetic influence, I would have painstakingly documented the evidence for the 'results of genetic research' column in Table 2. Our survey results indicate that the zeitgeist has changed sufficiently so that it is no longer necessary to do that. Most people accept that DNA matters for psychological traits, even though they underestimate its influence.

I hope my reading of the zeitgeist is correct because, otherwise, there would be a huge amount of research to review, many tens of thousands of studies, with more than 20,000 papers published during the past five years alone. It would be boring to condense this research here because the bottom-line message is similar for all areas of psychology. As you can see in Table 2, psychological traits are all substantially heritable, about 50 per cent on average.

Heritability is so ubiquitous that this has been called *the first law of behavioural genetics*: All psychological traits show significant and substantial genetic influence.

The results from our survey suggest that it is no longer necessary to convince most people that DNA matters for human individuality. Rather than reviewing the mountain of evidence that supports the 'results of genetic research' column in Table 2, we will examine in the next chapter the methods and some examples of the results that led to the first law of behavioural genetics.

The first part of *Blueprint* presents some of the biggest findings in psychology, findings that go far beyond estimating heritability. These discoveries came from adding genetics to mainstream psychological research, which had previously ignored genetics. By disentangling the effects of nature and nurture rather than assuming that nurture alone was responsible for who we are, this research produced startling results that suggest a completely different way to think about the roles of nature, nurture and their interplay in making us who we are.

2

How do we know that DNA makes us who we are?

In cognitive psychology, anecdotes and thought experiments can get basic ideas across, like the mistakes we often make when we think intuitively. In neuroscience, pictures of bits of brain lighting up suffice to light up ideas. Evolutionary psychology is also easy to describe, because its evidence rests on average differences between species. What is difficult about describing genetic influences in psychology is that genetics is not about how we all think or how our brains work in general or what we are like as a species. Genetics is about differences between individuals, rather than between groups. It is the essence of our individuality.

To describe the genetic origins of individual differences, anecdotes are not enough and thought experiments are not possible. Understanding the basis for the estimates of genetic influence in the previous chapter requires a grasp of the methods and analyses used to come up with these estimates. This requires some statistics, too, the statistics of individual differences.

In this chapter, I use individual differences in body weight to illustrate the methods of behavioural genetics for three reasons. First, although weight is a physical characteristic, it is a major area of research in health psychology. Weight is the result of behaviour – what we eat and how much we eat and how much we exercise – and psychology is the science of behaviour. In many ways, the obesity epidemic is a psychological problem.

Second, as we saw in the survey in the previous chapter, people think that weight is much less heritable than it is (40 per cent versus 70 per cent). I hope that this makes the evidence for its 70 per cent heritability more interesting. Third, no one questions that you can measure weight accurately. In contrast, measurement of psychological traits is less clear cut. For example, personality traits are usually assessed using self-report questions and psychopathology is diagnosed on the basis of interviews.

Weight raises all the issues relevant to understanding the origins of psychological traits. The starting point for genetic analysis is familial resemblance – does the trait run in families? For weight, the resemblance is strong enough that you can see it yourself if you look across families you know. Thin people are likely to have parents and siblings who are thinner than most people in the population. If weight did not run in families, genetics could not be important.

Weight can run in families for reasons of nature (genetics) or nurture (environment). For a century, genetic research has relied on two methods to disentangle nature and nurture: the adoption method and the twin method. The two methods have different assumptions, strengths and weaknesses. Despite the great differences in the two methods, the results of adoption and twin studies converge on the same conclusion about the importance of inherited DNA differences in the origins of psychological traits.

A SOCIAL EXPERIMENT: ADOPTION

One way to disentangle nature and nurture is to find relatives who share nature but not nurture in order to test the power of genetics. Adoption is like a social experiment that does just this. We can see how similar children are to their biological, or 'genetic', parents when the children are adopted away at birth. These parents share nature but not nurture with their children. If nature is why weight runs in families, adopted children should resemble their genetic parents, not their adoptive parents.

Adoption studies also provide a direct test of nurture. If nurture is why weight runs in families, adopted children should resemble their adoptive parents, who are their 'environmental' parents. Just like parents who rear their genetic children, adoptive parents provide their children's family environment, including the food they eat, and model healthy or unhealthy lifestyles.

Nonetheless, parents and their children differ by at least two decades in age and they grow up in different environments. Therefore,
an even better test of the influence of family environment is to study
'environmental' siblings. About a third of adoptive families adopt
two children. These children have different biological parents and are
not genetically related but they grow up in the same family. If nurture
explains individual differences in weight, adoptive siblings should be
just as similar as siblings who share both nature and nurture.

At the beginning of my career I had the chance to conduct an adoption study at a time when adoption was much more common than it
is today. In 1974, after finishing my PhD at the University of Texas at
Austin, I got my dream job at the University of Colorado at Boulder
with a joint appointment in the Department of Psychology and the
Institute for Behavioral Genetics, the only institute of its kind in the
world. I decided to create a long-term longitudinal adoption study of
psychological development. It was considered a classically bad idea
for a new assistant professor to begin such a long-term project because
it would not pay off soon enough to ensure that they kept the job and
would be promoted. But I am an incurable optimist.

The adoption design is particularly powerful in disentangling
the influence of nature and nurture because it can include 'genetic'
parents, 'environmental' parents and 'genetic-plus-environmental'
parents. 'Genetic' parents are birth parents of adopted-away children, and 'environmental' parents are the adoptive parents of these
children. 'Genetic-plus-environmental parents' refers to the usual
situation in which parents share both nature and nurture with their
children. This design enables powerful estimates of genetic and environmental influence.

Adoption was at its peak in the early 1970s in the US. The swing-
ing sixties swung into a sexual revolution. The percentage of babies
born to unmarried women tripled from less than 4 per cent before
1960 to over 15 per cent by the 1970s. Although the birth-control pill
was approved in 1960 by the US Food and Drug Administration
and became widely used by married women, young single women
did not take it up until the mid-1970s. Abortion was prohibited and
an unmarried woman raising a child by herself was frowned on. It
was not until 1973 that the US Supreme Court *Roe v. Wade* decision

legalized abortion during the first trimester of pregnancy, and it took several years before legalized abortion became available.

In the 1970s young women pregnant 'out of wedlock', especially religious women, often went away to have their babies, staying in 'homes for unwed mothers' and then relinquishing their babies for adoption. The adopted-away children didn't see their birth mothers after the first week of life and adoption records were kept secret. Now there are many fewer adopted children and most adoptions are 'open', allowing contact between birth parents and adoptive parents.

During my first months in Boulder I identified two private religiously affiliated adoption agencies in Denver which arranged adoptions for several hundred newborn babies each year. To my surprise, the adoption agencies readily agreed to collaborate with me in this research.

Together we solved several problems. The major issue was maintaining the anonymity and confidentiality of the mothers and their children. These young women, mostly teenagers (their average age was nineteen), had left their own homes, friends and family to give birth without anyone knowing. They wanted nothing more than to return to their lives unscathed by their motherhood. We worked out a system in which the pregnant women provided no identifying information so that there would be no way to have further contact with them.

Several dozen of these young women lived together during the second half of their pregnancies in special-care homes managed by the adoption agencies. My plan was to test them in groups in their respective care homes. I tried to get as much information as I could on them during the agreed three-hour visit because our agreement was that I would have no further contact with them. The measures included cognitive tests and questionnaires about personality, interests and talents, and psychopathology. I also collected information about education and occupation, smoking and alcohol consumption, and height and weight.

I wanted to give the adoptive parents of these children the same battery of tests. And I wanted to visit the adoptive parents in their homes to study the development of their children. The adoption agencies encouraged adoptive parents to be open about adoption, especially with their children. Because they did not treat adoption as secretive, I was able to explain the project to groups of potential adoptive parents and found that most were eager to participate. I think this eagerness reflected their desire to learn about children and their development. Although many more newborns were available for adoption in the early 1970s than are now, it was still not easy to adopt a child. For example, adoptive parents had to provide evidence that they were infertile. They were interviewed extensively about their reasons for wanting to adopt a child, and they had to agree to visits by a social worker to assess the suitability of their home. The average time from first contact with the agency to placement of a child was three years.

Because the adoption agencies were religious, not-for-profit charities, they did not select adoptive parents on the basis of their wealth, although they did require that at least one parent was a practising Christian. The adoptive parents were reasonably representative of US families with children in terms of education and occupational status.

For two years most of my weekends were spent driving thirty miles from Boulder to Denver to conduct tests with groups of unwed mothers. It was easy to collect data from this captive audience because their main problem was boredom while living in the communal homes for several months. Almost all the mothers agreed to participate.

Genetic influence of parents on their children's development can be estimated directly from the resemblance of the 'genetic' parents and their adopted-away children. The flipside of the adoption design provides a direct estimate of the influence of 'environmental' parents – adoptive parents and their adopted children. After I received funding that allowed me to employ researchers to help with the testing, I obtained a matched sample of 'control' parents – parents who gave birth and reared their own child. These are 'genetic-plus-environmental' parents. All parents agreed to take the same assessment battery as the birth mothers.

My goal was to study 250 adoptive families and 250 matched control families in their homes yearly during infancy and early childhood. A third of the adoptive families adopted a second child and I also wanted to study these children, as well as siblings in the control families. I was particularly keen, for the first time in an adoption study, to assess the family environment using questionnaires, interviews and observations, including videotaped observations of interactions between parents and their children.

The study, called the Colorado Adoption Project (CAP), did not, however, end in early childhood because the value of the study increased with each wave of assessment. The children were studied in the laboratory at the ages of seven, twelve and sixteen, with telephone interviews in intervening years. At age sixteen, more than 90 per cent of the CAP children completed the same assessments their parents had completed sixteen years earlier. Parents and home environments were assessed through these years with questionnaires and telephone interviews. The study continues today, with the children now in their forties.

The results have been described in four books and in hundreds of research articles. CAP added to the evidence in support of the first law of behavioural genetics, that psychological traits show significant and substantial genetic influence. For example, even in childhood, we demonstrated genetic influence on intelligence, on specific cognitive abilities including verbal ability, spatial ability, on different kinds of memory, such as recalling names for faces, and on reading ability as early as seven years of age. Genetic influence was also found for infant temperament, as rated by observers, especially shyness. Ratings given by teachers of temperament indicated that it was highly heritable in adolescence. Behaviour problems also showed significant genetic influence, such as parent and teacher ratings of attention problems, as well as self-reported loneliness.

However, CAP's most important contribution was discovering some of the 'big findings' described in the following chapters. For example, it was the first study to report genetic influence on measures of the environment. How can environmental measures show genetic influence? You will find the answer in the next chapter.

A BIOLOGICAL EXPERIMENT: TWINS

If adoption is a social experiment separating the effects of nature and nurture, twins are a biological experiment. Where you can most clearly see heredity in action is identical twins. Identical twins come from the same fertilized egg, or zygote. This is why they have the same inherited DNA and why, in scientific terminology, they are called monozygotic twins (MZ). About one in 350 people is an identical twin, so chances are you personally know at least one pair of MZ twins.

If you don't know MZ twins personally, you have probably heard of famous MZ twin pairs, such as Cameron and Tyler Winklevoss, the internet entrepreneurs who created a social networking site at Harvard which they claimed was the inspiration for Facebook. You may also have heard of the American football players Ronde and Tiki Barber. The infamous 1950s East End criminals Ronnie and Reggie Kray were MZ twins. So are Ashley and Mary-Kate Olsen. They claim that they are not actually MZ twins despite looking very similar, a claim that could be easily proven with a DNA test. If they show any inherited DNA differences, they cannot be MZ twins.

If weight were 100 per cent heritable, MZ twins would have the same weight. As with other family members, the similarity in weight for MZ twins could be due to nurture as well as nature. The most dramatic test of genetic influence is to study MZ twins separated by adoption early in life. They share nature completely but do not share nurture at all, so their similarity is a direct test of genetic influence.

MZ twins reared apart are of course extremely rare. Only a few hundred pairs have been studied worldwide. These cases have produced some amazing examples of similarity. One of the first pairs studied extensively are the 'Jim twins', who were born in Ohio in the late 1930s. They were adopted separately at the age of four weeks by different couples who did not know that the child they adopted was one half of a twin pair. They are famous because, when they were reunited for the first time in 1979 at the age of thirty-nine, they reported some striking similarities. For example, both Jims did poorly in spelling and well in mathematics. They had similar hobbies in carpentry and mechanical drawing. They both began suffering from tension headaches at the age of eighteen, gained 5 kilograms at the same age, and are both 183 centimetres tall and weigh 82 kilograms.

But these are anecdotes, and the plural of anecdote is not data. Even though there are not many pairs of MZ twins reared apart, their results support other genetic research in pointing to substantial genetic influence. In general, MZ twins reared apart are almost as similar as MZ twins reared together, indicating that what makes them so similar is nature, not nurture.

The most widely used method to separate the effects of nature and nurture is to study twins reared together. Twins are a gift to science because there are two types of twins, not just MZ twins. About 1 per cent of all births are twins. One-third of these are MZ twins. The rest are called dizygotic (DZ), or fraternal, twins because they come from two eggs that are fertilized at the same time. Like any brother and sister, DZ twins are 50 per cent similar genetically.

Both MZ and DZ twins grow up in the same womb and, generally, in the same home. So, if nature is important for a trait, you have to predict that MZ twins will be more similar than DZ twins. If individual differences for a trait are caused entirely by inherited DNA differences, identical twins would correlate 1.0 for the trait, and fraternal twins would correlate 0.5. If genetic differences are not important, identical twins would be no more similar than fraternal twins.

In 1994 I received an exciting offer to move to London to help create an interdisciplinary research centre. The goal of the centre was to bring together genetic and environmental strategies to study the interplay between genes and environment in psychological development. This explains the centre's seventeen-syllable name – Social, Genetic and Developmental Psychiatry Centre – and it continues to flourish at the Institute of Psychiatry, Psychology and Neuroscience, King's College London, where I still work.

This move gave me the opportunity to begin a new long-term longitudinal study, this time a study of twins. I wanted to create a huge national twin study that would have the power to tease apart the effects of nature and nurture in development. The only way to do this systematically is to identify twins from birth records. Although I started a twin study in Colorado that focused on infancy, it would be difficult to create a national twin study in the US because birth records are controlled separately by each state. In the UK I was lucky because birth records had just been computerized, in 1993, at which time the birth records also began to record for the first time whether there was a twin.

About 7,500 pairs of twins are born each year in the UK. I aimed to invite parents of twins born in 1994, 1995 and 1996, which would include more than 20,000 pairs of twins. I wanted to study the twins' psychological development from birth and to follow them through infancy, childhood, adolescence and adulthood to explore how genetic and environmental influences change from age to age. I called the study the Twins Early Development Study (TEDS).

TEDS got off to a roaring start. Parents of twins participate in research twice as much as other parents because they understand that twins are special and that studies of them can advance the cause of science. In TEDS, more than 16,000 families of one-year-old twins agreed to take part. I find this particularly impressive because having twins is more than twice the work of having a single child. These parents had their hands full, yet they readily agreed to contribute to the research.

The twin method is based on comparing identical and fraternal twins. How can you tell whether a twin pair is identical or fraternal? Because identical twins are genetically identical, they are very similar for all highly heritable characteristics, such as height, eye colour, hair colour and general looks. They are difficult to tell apart, sometimes to their annoyance (being confused for their twin) and often to their amusement (intentionally confusing others). Just asking a single question provides more than 90 per cent accuracy in deciding whether a twin pair is identical: Are they as similar as two peas in a pod?

Figure 1 shows how similar identical twins are. Rosa and Marge are identical twins who have participated in TEDS since they were two years old. Rosa is now a PhD student doing her doctoral research on TEDS. Marge is a PhD student in anthropology. In contrast, fraternal twins are no more similar than any sisters and brothers, as illustrated by the TEDS twin sisters in the lower half of Figure 1. Half of fraternal twins are opposite-sex twins. Because identical twins are always of the same sex, same-sex fraternal twins provide a better comparison group.

The ultimate test is DNA. Identical twins have identical DNA sequence, but fraternal twins show only 50 per cent similarity for DNA differences. So, if a twin pair shows DNA differences, they cannot be identical twins. This is why I said earlier that the issue of

Figure 1 Identical and fraternal twins

whether the Olsen twins are MZ twins could easily be resolved with a DNA test. TEDS has obtained DNA for more than 12,000 twins, which has achieved much more than verifying whether twins are MZ or DZ. It has put TEDS at the forefront of the DNA revolution.

The TEDS families were invited to take part in studies when the twins were aged two, three, four, seven, eight, nine, ten, twelve, fourteen and sixteen. We are now studying the twins again as they emerge into adulthood at the age of twenty-one. Unlike CAP, which had just 500 families, it was not possible financially to visit the thousands of

TEDS twins in their homes. Necessity was the mother of invention and we created new ways to assess children's development remotely. When the children were aged two, three and four, we enlisted the twins' parents as testers to gauge the twins' cognitive and language development. When they were seven, we devised measures of cognitive ability to administer to the twins over the telephone. By the time the TEDS twins were ten, access to the internet in UK homes was sufficient for us to administer cognitive tests online. Since then, all our assessments have been online.

We also created web-based tests of the cognitive skills taught in school, especially reading and mathematics. In addition, we have been able to use data on the TEDS twins from the UK National Pupil Database, which includes standardized school achievement data on English, mathematics and sciences for all children at the ages of seven, eleven and sixteen.

Although cognitive and language development was a focus of TEDS, we also collected questionnaire data from parents, teachers and eventually the twins themselves about psychological problems, health and home and school environments.

Altogether, the TEDS data set consists of 55 million items of data collected from parents, teachers and twins over twenty years. TEDS findings have been reported in more than 300 scientific papers and in 30 PhD dissertations. Like CAP, TEDS has shown that many traits (some of them in addition to those investigated in CAP) obey the first law of behavioural genetics. For example, in the cognitive domain, we found that how well children do at school in all subjects, from humanities to sciences, is substantially heritable. We also found that components of reading (for example, phonetics) and of language (for example, fluency) are highly heritable. For the first time, we showed that individual differences in learning a second language are highly heritable. We looked in depth at aspects of spatial ability, such as navigating from a map, with results again showing ubiquitous heritability.

In the realm of personality and psychopathology, we also investigated traits beyond those mentioned in the previous chapter. For example, we found substantial heritability in childhood for lack of empathy and disregard for others, known as callous-unemotional traits and thought to be an early sign of psychopathy. High heritability also emerged for symptoms of hyperactivity and inattention, which are components of attention deficit hyperactivity disorder (ADHD). We studied many aspects of well-being such as life satisfaction and happiness, again with similar results showing substantial heritability.

Adoption studies such as CAP and twin studies such as TEDS have different strengths and weaknesses for estimating genetic influence. Despite these differences, twin and adoption studies converge on the same conclusion that genetic influence on psychological traits is substantial. The first law of behavioural genetics is so well documented that what is interesting now is to use adoption and twin studies to go beyond estimating heritability.

Like CAP, the most important contribution of TEDS is its role in discovering the 'big findings' described in the following chapters. For example, TEDS took the lead in showing that what we call disorders are not genetically distinct from the normal range of variation. Although it might not sound very exciting, this finding has far-reaching implications for clinical psychology because it means that there are no disorders, that the 'abnormal is normal' (which is the title of one of the chapters to come).

Crucially, TEDS has been at the cutting edge of the DNA revolution, which is the focus of the second part of *Blueprint*.

Adoption and twin designs make clear predictions about what to expect if inherited DNA differences matter for individual differences in weight. For example, adopted children should resemble their genetic parents rather than their environmental parents. MZ twins should be more similar than DZ twins.

Data from adoption and twin studies can be used to ask whether there is any statistically significant evidence for genetic influence. But these data can also be used to estimate how much inherited DNA differences matter. It doesn't matter much whether DNA differences account for 40 per cent versus 50 per cent of individual differences in weight. But it matters whether DNA differences account for 40 per cent, as people rated it in my survey, or 70 per cent, which is what the research says. If the answer is 70 per cent, it means that most of the difference in weight between people is due to DNA differences, which has personal and policy implications, which I will discuss later.

To explain the estimate of 70 per cent, we need the statistics of individual differences. There are two basic statistics of individuality: variance and covariance. These are crucial not just for understanding genetics but also for interpreting all scientific data on individuality.

Variance is a statistic that describes the extent to which people differ, whereas covariance indexes similarity. Most people are more familiar with the term 'correlation', which describes the relationship between two traits. A more scientific way of explaining this is that correlation indicates the proportion of the variance that covaries. A correlation of 0.0 means that there is no similarity between the two traits, whereas a correlation of 1.0 means perfect resemblance.

To take an example, what do you think the correlation is between weight and height? Obviously, taller people weigh more, so the correlation is not zero. But how strong is the relationship? A correlation of 0.1 is small, a correlation of 0.3 is moderate, and a correlation of 0.5 is substantial. In fact, weight and height correlate 0.6. That's all you really need to know about statistics in order to make sense of genetic data. But if you would like to know more, in the Notes section at the end of the book I describe the statistics of individual differences in greater detail, using the correlation between weight and height as an example.

In genetics, the correlation is used to assess the association between two family members – two members of a twin pair, for example. In other words, instead of correlating traits such as height and weight in the same individuals, we correlate a trait for one member of a twin pair with the same trait in the other twin. The twin correlation indicates how similar the twins are. As before, a correlation of 0.0 means that the twins are not at all similar, and a correlation of 1.0 means that they are totally similar.

Figure 2 shows a scatterplot for weight between one member of a twin pair and the twin's partner, or co-twin, using data from TEDS. The first scatterplot is for 600 pairs of MZ twins and the second is for 600 pairs of same-sex DZ twins. DZ twins can be same sex or opposite sex, but because MZ twins are always of the same sex, a better comparison group is same-sex DZ twins.

Figure 2 Scatterplot showing MZ and DZ twin correlations for weight in 16-year-olds. The MZ correlation (*top*) is 0.84 and the DZ correlation (*bottom*) is 0.55.

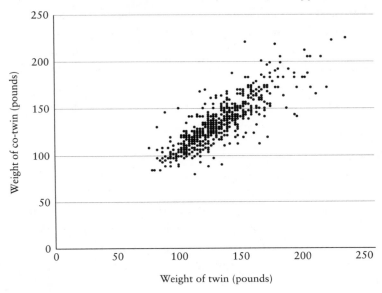

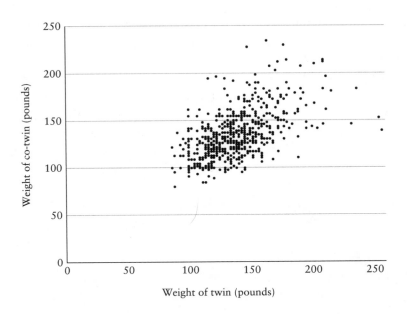

The scatterplots show that the correlation for MZ twins is greater than the correlation for DZ twins. The scatterplot is less scattered for MZ twins than for DZ twins. In other words, one twin's weight is a stronger predictor of the co-twin's weight for MZ twins than for DZ twins. The actual twin correlations for these TEDS data are 0.84 for MZ twins and 0.55 for DZ twins. The correlation of 0.84 for MZ twins is almost the same as the correlation between the same individuals measured twice a year apart. In contrast, the correlation for fraternal twins is much lower: 0.55. The fact that the MZ twin correlation is greater than the DZ twin correlation suggests genetic influence.

The difference between the MZ and DZ correlations can be used to estimate heritability. Heritability is central to this book because it indicates the extent to which DNA makes us who we are.

As described earlier, the most straightforward estimate of heritability comes from the correlation for MZ twins reared apart. Their correlation directly estimates heritability. If the correlation for MZ twins reared apart is 0, heritability is 0, whereas a correlation of 1.0 indicates heritability of 100 per cent.

Although MZ twins reared apart are extremely rare, results for several hundred such pairs have been reported. A well-known study in the US is the Minnesota Study of Twins Reared Apart, which comprised fifty-six pairs of MZ twins reared apart and included the 'Jim twins' mentioned earlier. Their correlation for weight was 0.73. I was involved in a study in Sweden that systematically identified twins from birth records and found more than a hundred pairs of MZ twins reared apart. Most of these twins were elderly, born in the early twentieth century. The reason for their separation was an economic depression in Swedish agrarian society at the time, coupled with a high risk of maternal death during twin birth. This resulted in many twins being put up for adoption and adopted separately early in life. These reared-apart twins became participants in our Swedish Adoption/ Twin Study of Aging. Their correlation for weight was also 0.73.

Across all studies of MZ twins reared apart, the twin correlation for weight is 0.75. This indicates that 75 per cent of the differences between people in weight (variance) is shared (covariance) by these pairs of genetically identical individuals who did not grow up in the same family environment. For this reason, the correlation between identical twins reared apart is a simple, direct estimate of heritability: the extent to which differences in weight between individuals can be accounted for by inherited DNA differences.

Most heritability estimates come from the classic twin design that compares correlations for MZ and DZ twins who were reared together, as in TEDS. Suppose the correlations for MZ and DZ twins were the same. This means that the twofold greater genetic similarity of MZ twins does not make them more similar than DZ twins. We would have to conclude that DNA differences are not important – heritability is 0 per cent. Heritability is 100 per cent if the correlations for MZ and DZ twins completely reflect their genetic similarity – 1.0 for MZ twins and 0.5 for DZ twins.

In TEDS, the MZ correlation for weight is 0.84, whereas the DZ correlation in TEDS is 0.55. Because DZ twins are only half as similar genetically as MZ twins, the difference in correlations (0.84 versus 0.55) estimates half the heritability of weight. Doubling this difference in correlations puts heritability as 58 per cent.

The heritability estimate from TEDS is about 60 per cent, but the estimate from all research is 80 per cent. Why do these two estimates of heritability differ? The answer is an example of another of the 'big findings' of genetic research: heritability increases during development. Twins in TEDS are adolescents, but most other twin studies involve adults. In an analysis of forty-five twin studies, the heritability of weight increases from about 40 per cent in early childhood to about 60 per cent in adolescence to about 80 per cent in adulthood. The heritability estimate of 60 per cent from the adolescent twins in TEDS is just what would be expected. When we study the TEDS twins later in adulthood, we will expect to find a heritability estimate closer to 80 per cent.

Adoption studies also converge on the conclusion that the heritability of weight is substantial. CAP results for body weight illustrate how the adoption study works. Weight definitely runs in families. The correlation between weight of parents and children is about 0.3 in 'control' families in which parents and their children share both nature and nurture.

Is this similarity in weight between parents and their young children a sign of nature or nurture? The CAP results provide a clear and consistent answer. The weight of adopted children does not correlate with the weight of their adoptive parents. Their correlation is just about 0. This means that dietary and lifestyle differences of adoptive parents are not at all related to the weight of their adopted children. Similarly, siblings correlate about 0.3 for weight, but when two unrelated children are adopted into the same home their correlation for weight is near 0. Growing up in the same family does not make children similar in weight unless the children share genes.

Just as stunning is the CAP finding that the correlation between these same adopted children and their birth mothers is about 0.3, the same as the parent–offspring correlation in control families. Even though these children were adopted away from their mothers at birth, their similarity in weight to their birth mother is the same as it is for children who are reared by their birth mother.

These adoption data all suggest genetic influence. The data can also be used to answer the question of 'how much' influence, that is, to estimate heritability. Because parents and offspring and siblings are only 50 per cent similar genetically, their correlation estimates only half the genetic influence on weight. So, the correlation of 0.3 between adopted children and their birth parents is doubled to estimate the heritability of weight as 60 per cent.

This evidence for the importance of nature can obscure a crucial finding about nurture from adoption studies. Isn't it astonishing that the correlations are near 0 between adoptive parents and their adopted children and between adoptive siblings? Even though adoptive parents put the food on the table, their adopted children are not at all similar to them in weight. Similarly, even though adoptive siblings grow up together in the same family, sharing the same parents, the same food and the same lifestyle, they are not at all similar in weight.

These results for adoptive parents and their adopted children and for adoptive siblings indicate that weight runs in families for reasons of nature, not nurture. The environment is important. The heritability of 60 per cent implies that environmental forces account for 40 per cent of the differences in weight. But nurture – that is, sharing a family environment – has little effect on individual differences in

28

weight. This is another of the big findings from genetic research, which, as I will discuss later, has been found to apply not just to weight but to all psychological traits. This is the topic of Chapter 7.

Putting all of these twin and adoption data together using a technique called model-fitting comes up with an estimate of about 70 per cent for heritability of weight. This overall estimate averages out issues such as increasing heritability of weight over time. It also glosses over several nuances about differences in twin and adoption designs that are intriguing to behavioural geneticists but probably not of much interest to most people.

One nuance of more general interest is group differences. The overall estimate of 70 per cent heritability might mask differences between certain groups. For example, does heritability differ for males and females? The answer is 'no'. Does heritability differ in different populations? The answer is 'not much'. Most studies have been conducted in developed countries so it is possible that developing countries show different results. Within developed countries, there is some recent evidence that heritability of weight may be higher in richer countries with richer diets. Perhaps easy access to high-energy foods encourages people with a greater genetic propensity to gain weight.

The point is that these very different designs – twin and adoption studies – converge on a simple but powerful conclusion: most of the differences between people in weight can be explained by inherited differences in DNA.

Thousands of studies have used these twin and adoption methods to explore the extent to which DNA matters for thousands of complex traits throughout biological and medical sciences, including just about anything that can be measured, from cells to systems, such as structural and functional measures of brain, heart, lungs, stomach, muscles and skin. A recent review of twin studies looked at 18,000 traits in 2,700 publications that included nearly 15 million twin pairs. Across all traits, the average heritability was 50 per cent. Although body weight is more heritable than most traits, all psychological traits show substantial genetic influence. This is the evidence for the first law of behavioural genetics.

Discovering that DNA matters so much in psychology is a fundamental achievement of behavioural genetics. This first law of behavioural genetics is so well established that it is no longer interesting to show that some new trait is heritable, because all traits are heritable. Behavioural genetics has moved beyond heritability to ask novel questions. These questions include developmental change and continuity, the links between traits, and the interface between nature and nurture. This research has produced some of the most important findings in psychology, and I will explore them in the following chapters.

It cannot be overemphasized that genetic effects on psychological traits are not only statistically significant, they are massive in terms of how much variance they explain. The size of an effect – *effect size* – is the critical issue in interpreting research on individuality. Statistically significant findings may not be significant in any real-world sense if their effect size is negligible. Statistical significance depends on sample size – with a large sample, a tiny effect size can be highly significant statistically. What's really important is effect size, that is, variance explained.

Few effect sizes in psychology are greater than 5 per cent. As one of countless examples, much is made of the differences between boys and girls, for instance in school achievement. Although this difference is highly significant statistically, the question that needs to be asked is about effect size: How much do boys and girls actually differ in school achievement? The answer is that sex differences account for less than 1 per cent of the variance. In other words, if all you know about a child is whether the child is a boy or a girl, you know practically nothing about their propensity to achieve at school.

This is why it is incredible to find that 50 per cent of the differences between people in psychological traits are caused by genetic differences between them. The heritability effect size of 50 per cent is off the scale of effect sizes in psychology. As a rule of thumb, we can classify effect sizes as small, medium and large. Explaining 1 per cent of the variance is a small effect, an effect so small that you can't see it without statistics. Most effect sizes in psychology are small, as in the example of sex differences in school achievement. Another example related to school achievement is classroom size – it is widely assumed that children learn more in classrooms containing fewer children. The correlation between the number of pupils in a class and educational achievement is highly significant statistically because it is based on huge sample sizes. But the effect size is only 1 per cent.

A medium effect explaining 10 per cent of the variance can be seen with the naked eye, although you might have to squint to see it. For example, parental educational attainment explains almost 10 per cent of the variance in their children's educational attainment. Among people you know you can see that children are more likely to go to university when their parents are university educated. As we shall see, this correlation is mostly down to nature, not to nurture, as you might assume.

A large effect explains 25 per cent of the variance, an effect so large you would stumble over it in the dark. There are very few large effect sizes in psychology. One example is that general intelligence accounts for about 25 per cent of the variance in educational achievement. On this scale from small (1%) to medium (10%) to large (25%) effect sizes, heritability of 50 per cent is literally way off the scale. Inherited DNA differences are by far the most important systematic force in making us who we are.

3

The nature of nurture

Even before the DNA revolution behavioural genetics produced some of the biggest findings in all of psychology – 'big' in the sense of how much they shape who we are and also in the sense of their importance for understanding our society and ourselves. In this book I focus on the five most significant findings of the past few decades, which we will explore in greater detail in the following chapters.

There are three things about these findings that are especially important. First, they are counterintuitive. Findings that confirm received wisdom can be important, but findings that clash with what is intuitively obvious are more likely to lead to breakthroughs.

The second important thing about these findings is that two of the five are about the environment. Genetic research has told us as much about the environment as it has about genetics. At the most basic level, genetics provides the best evidence we have for the importance of the environment independent of genetics. That is, heritabilities are never even close to 100 per cent, which proves that the environment is important. Traditionally, environmental research has ignored genetics and thus could not untangle the threads of nature and nurture. Genetic research has made fundamental discoveries about the environment because it takes genetics into account when studying the environment. This research has fundamentally changed the way we think about nurture and its intersection with nature.

The third thing is that these findings are solid – they have been replicated many times and in many ways. You might think that replication could be taken for granted in science. But there is currently a crisis in science about failures to replicate. It began in 2005 with a paper with the shocking title 'Why Most Published Research Findings are False'.

This is such an important issue in science today that I want to preface these chapters on big findings from behavioural genetics by describing this crisis and considering the reasons why big findings from behavioural genetic findings replicate so robustly.

The bottom line of science is replication, that is, results need to be reliable, in the sense that they can be replicated. The current crisis is that the results of many studies, including classic studies that are the backbone of textbooks, do not replicate, creating gaping cracks in the bedrock of science. Failures to replicate are popping up all over science, including medicine, pharmacology and neuroscience as well as psychology. In relation to psychology, an influential paper in the journal *Science* reported that more than half of 100 studies in top journals failed to replicate.

Much has been written about the causes of this crisis. Outright fraud occurs, but it is rare. One general factor is the hypercompetitive culture for publishing novel results in the best journals, which increases the risk for what can only be called cheating. This cheating is unconscious perhaps, but it is cheating nonetheless, for example, when scientists select results that tell the best story, sweeping inconsistencies under the carpet. As the physicist Richard Feynman said, 'The first principle is that you must not fool yourself – and you are the easiest person to fool.'

A specific source of cheating is called *chasing probability (P) values*. Although this topic sounds esoteric, it is an important insight into how science is supposed to work. A *P* value of 5 per cent is a convention used in science as a threshold for concluding that results of a study are statistically significant. When a scientist says results are significant, this only means statistically significant, not significant in the usual sense of the word. Reaching a *P* value of 5 per cent means that if you did the same study one hundred times you would find a similar result ninety-five times. A *P* value of 5 per cent does not mean that a finding is true. It means that five times out of one hundred tries you would not find the same 'significant' result, which are called *false positive results*. If you find a result significant at the *P* value of 5 per cent, it could be one of those false positive results.

Because scientific journals only publish results that are statistically significant, false positive findings are expected 5 per cent of the time.

However, false positive findings appear in published papers much more than 5 per cent of the time, for two main reasons. First, these published results are often novel and interesting findings – and thus more publishable – precisely because they are not true. Second, the situation edges closer to cheating when scientists 'chase P values'. For example, they might look at their data in different ways – for example, using different types of analyses – and choose to write about the results that reach the P value of 5 per cent. But chasing P values in this way chases the validity of statistical tests right out of the window.

Many other causes of the replication crisis have been discussed. Dozens of papers have also been written about how to fix these cracks in the foundation of science. For example, there is momentum to solve the problem of chasing P values by playing down statistical significance and focusing instead on how big the effect is. Effect size is the critical issue in interpreting research on individuality. Very often, statistically significant findings are not significant in any real-world sense because their effect size is negligible. Statistical significance depends on sample size and effect size. A tiny effect size will be statistically significant if the sample size is large enough. So, when you hear about a scientific finding, always ask about the size of the effect. It is not enough to know that the finding is statistically significant.

Behavioural genetic research is as vulnerable as other fields to the risks for reporting false positive results that fail to replicate. Nonetheless, the general finding that all psychological traits are substantially heritable and the five big findings described in the following chapters have been replicated many times. Why do findings in behavioural genetics replicate so strongly? The main reason for the robustness of behavioural genetic results is that genetic effect sizes are so big it is difficult to miss them if you look for them. Inherited DNA differences account for 30 to 60 per cent of the variance for most psychological traits. Few other findings in psychology account for 5 per cent of the variance.

Another reason seems paradoxical: behavioural genetics has been the most controversial topic in psychology during the twentieth century. The controversy and conflict surrounding behavioural genetics raised the bar for the quality and quantity of research needed to convince people of the importance of genetics. This has had the positive effect of motivating bigger and better studies. A single study was not enough. Robust replication across studies tipped the balance of opinion.

New methods that assess DNA differences directly are also beginning to confirm these findings that were based on twin and adoption studies. Replicating these findings using DNA alone will convince even more people that DNA matters. Twin and adoption studies are indirect and complicated. But it is difficult to doubt results based directly on DNA.

The DNA revolution matters much more than merely replicating results from twin and adoption studies. It is a game-changer for science and society. For the first time, inherited DNA differences across our entire genome of billions of DNA sequences can be used to predict psychological strengths and weaknesses for individuals, called *personal genomics*. After we explore the big findings from behavioural genetics and their implications, the second part of *Blueprint* will focus on the DNA revolution.

Genetics makes us rethink some of our basic assumptions about how the world around us shapes who we are – or doesn't. The best example is a topic I have called *the nature of nurture*, which leads to a new understanding of what the environment is and how it works.

When we think about nurture, images come to mind like parents cooing to and cuddling their babies. Freud thought that parenting is the essential ingredient in a child's development. He focused on specific aspects of parenting, including breastfeeding and toilet-training, and how they affect sexual identity. He wrote persuasively about clinical case studies that supported his ideas, but he provided no real data. When research was done to test his ideas, little support was found for them. The philosopher of science Karl Popper claimed that Freud's theories were presented in a form that made them impossible to disprove, which is the Popperian sin against the first commandment of science that theories be not just testable but falsifiable.

Since Freud, thousands of studies in the behavioural sciences have investigated other aspects of parenting, such as warmth and discipline, as environmental influences on children's development. It is important to remember that we are always talking about individual differences – for example, why some parents are more loving towards or more controlling of their children as compared to other parents. Developmental psychologists study differences in parenting in order to ask whether differences in parenting cause differences in children's outcomes. For example, do differences in parental warmth make a difference in how well adjusted their children are later in life?

As children go to school, they experience a new world of class-rooms and playgrounds full of other children, potential friends and foes. Teachers can be inspirational role models, classmates can be bullies. For adults, a huge area of environmental research involves life events, which includes crises like financial problems and relationship breakdowns.

Parenting and life events are archetypes of measures of the envir-onment that have been used in thousands of psychological studies. These measures are then correlated with psychological traits to inves-tigate the influence of the environment. How much parents read to their children is correlated with how well the children learn to read at school. Hanging out with bad peers is correlated with bad outcomes such as using drugs in adolescence. Breakdowns in relationships and other stressful life events are correlated with depression.

It seems reasonable to assume that these correlations between environmental measures and psychological outcomes are caused *environmentally*. For example, the correlation between how much parents read to their children and how well the children learn to read at school seems likely to be caused by how much parents read to their children. Hanging out with bad peers seems to cause bad adolescent outcomes. Stress seems to cause depression.

As reasonable as these causal interpretations appear to be, we should be wary of interpreting any correlation in terms of one thing causing the other. It is always possible to interpret these correlations in the opposite direction: the dictum that correlation does not imply causation. For example, rather than parents' reading to children causing differences in how well the children read at school, how much parents read to children might reflect how much children enjoy read-ing. In addition, it is possible that neither thing causes the other. A third factor might set up the correlation between them. A classic example is the correlation between the number of churches in cities and the amount of alcohol consumed. Religion does not drive you to drink, nor does drinking make you more religious. The correlation is caused by the size of cities: because larger cities have more people, they have more churches and greater consumption of alcohol. Once you control for this third factor, there is no association between the number of churches and the amount of alcohol consumed.

Genetics could be a 'third factor' that contributes to the correlation between parents' reading to their children and their children's reading ability at school. This is what I mean by the nature of nurture. Because parents and their children are related 50 per cent genetically, it is possible that genetics creates the correlation between parents who read to their children and children who are good at reading. The association could be phrased in a way that makes the possibility of genetic mediation more obvious: parents who like to read have children who like to read. Another entry point for genetics is that children who like to read or be read to might use their environment to feed their appetite for reading, for example, by asking their parents to read to them. In other words, parents might be responding to genetic differences between children in how much they enjoy reading.

What if we analysed environmental measures in a genetic design like a twin study? It seemed like a silly thing when I first did this in the 1980s because environmental measures should not show any genetic influence – after all, they are environmental measures. Or are they? This was how the nature of nurture phenomenon was first discovered.

One of the early examples of the nature of nurture was what psychologists call stressful life events. These are part of the routine ups and downs of life, such as relationship breakdowns, financial difficulties, problems at work, illnesses and injuries, and being robbed or assaulted.

People differ in how they respond to events like these. Measures of life events incorporate the effect of an event because people can experience the same event very differently. Despite all the research on life events, no one had ever asked if individual differences in these experiences were influenced by genetic differences. If life events are just a matter of bad luck, they should not show genetic influence.

In the first genetic analysis of stressful life events in 1990 we studied middle-aged twins from Sweden, twins reared apart as well as twins reared together, in a study called the Swedish Adoption/Twin Study of Aging (SATSA). We included a questionnaire called the Social Readjustment Rating Scale, which has been used in more than 5,000 studies as a measure of the environment and includes standard items such as changes in relationships, financial status and illness. In addition, because our twins were sixty years old on average, we used a version of the questionnaire that adds items relevant to later life such as retirement, loss of sexual ability or interest and death of spouse, siblings or friends.

We were surprised to find that identical twins were twice as similar as fraternal twins in their scores on the measure of life events (twin correlations of 0.30 and 0.15, respectively). The same pattern of results emerged for the twins who had been reared apart in different families. These twin correlations suggest that inherited DNA differences account for about 30 per cent of the differences between people. What's amazing about this is that stressful life events had been assumed to be completely environmental in origin but almost a third of its variance is genetic in origin.

How can stressful life events show genetic influence? The questionnaire used in this study combined perceptions of whether an event occurred and how you respond to the event. Genetic influence on personality can affect both these perceptions. People differ in what they are willing to call a serious illness or injury, financial difficulty or relationship breakdown. Personality is especially involved in how much they feel these events affected them. Optimists might see these experiences through rose-coloured glasses, while pessimists see them in shades of grey.

What about stressful events themselves, free of perception? Divorce is an example of an objective event and one of the most stressful life events for most people. The first genetic study of divorce caused a stir. In a study of 1,500 pairs of adult twins, concordance for divorce was much greater for identical than for fraternal twins (55 per cent versus 16 per cent), suggesting substantial genetic influence on divorce. *USA Today* called this study 'the epitome of asinine' because it seemed preposterous to conclude that divorce is influenced by genetic factors.

But is it 'the epitome of asinine' to think that the objective event of divorce could be influenced by our genetically rich differences in personality? To the contrary, I think it is unreasonable to assume that events like divorce are just things that happen to us, as if we have nothing to do with them.

I hope it is clear by now that, contrary to newspaper headlines at the time, this research is not saying that there is a 'divorce gene' that makes some people hard-wired to get divorced. Nor are there 'bad genes' that make some people poor prospects for stable marriages. Subsequent research has shown that certain personality traits account for a third of the genetic influence on divorce. Surprisingly, people are more likely to get divorced if they are joyful and engaged with life, emotional and impulsive. These are not bad aspects of personality – indeed, they might be the same good traits that make people desirable as marriage partners in the first place.

It has long been known that the offspring of divorced parents are more likely to get divorced themselves. Possible environmental explanations leap to mind, for example, living through their parents' divorce causes children to have relationship problems, or because they do not have good models for a stable relationship. However, a recent adoption study in Sweden showed that the link between divorce in parents and divorce in their children is forged genetically, not environmentally. For a sample of 20,000 adopted individuals, the likelihood of divorce was greater if their biological mother, who did not rear the individual, had later in life become divorced than if the adoptive parents who reared them had become divorced.

The heritability of divorce is about 40 per cent across studies. This is a long way from 100 per cent, meaning that non-genetic factors are also important. However, the major systematic factor affecting divorce is genetics. In contrast, no environmental predictors of divorce have been identified in research after controlling for genetics. Controlling for genetics is crucial, as seen in the Swedish adoption study. Parental divorce is the best predictor of children's divorce but this association, easily interpreted as environmental, is actually due to genetics.

So, divorce doesn't just happen by chance. We make or break our relationships. We are not just passive bystanders at the whim of events 'out there'. As always, genetic influence means just that – *influence*, not hard-wired genetic determinism. There are no *schlimazel* (Yiddish for 'crooked luck') genes that attract life's pies in the face.

It's not just life events. Calling any measure 'environmental' does not make it a measure of the environment. Genetic studies of environmental measures have found significant heritability for most measures of the 'environment' – parenting, peer groups, social support and even how much time children spend watching television.

Children's television viewing is a quintessential measure of the environment, which, by the 1980s, had been used in more than 2,000 studies exploring its effect on children's development. None of these studies questioned the assumption that how much television children watched was a measure of the environment. The basic message was that the one-eyed monster was bad for children – making them do less well at school and making them more aggressive and less attentive. Correlations between television viewing and children's development were always interpreted in this way, as being caused environmentally.

At that time, in the early 1980s, I also assumed that differences in how much television children watched was a matter of the environment because I thought parents were in charge of how much television their children watched. Although my wife and I were generally permissive, we also believed television was bad for children and we controlled how much television our two young sons watched.

If parents are in charge of their children's viewing time, this could diminish the role of genetics in their viewing time. But as I read more about it I was surprised to learn that most parents back then put no restrictions on the amount of time their children watched television. How much children watched television was up to the children, which leaves the door open for genetic differences between children to shape how much television they watch.

For these reasons, I decided to study children's television viewing in the Colorado Adoption Project. When we visited the 500 adoptive and non-adoptive families as the children turned three, four and five years of age, we interviewed parents for ten minutes about how much television their children watched and what programmes they watched.

It took almost five years to collect the data at these three ages.

When I finally analysed the results for television viewing I expected to find little evidence for genetic influence. I first calculated the correlations for the non-adoptive siblings, who share both genes and family environment. The correlations at the three ages were about 0.50, indicating that non-adoptive siblings watched similar amounts of television. This is not surprising because siblings often watched television together, especially in those days, when most families just had one television. However, I was stunned when I looked at the correlations for adoptive siblings because they were consistently about half the size of the correlations for non-adoptive siblings. Because adoptive siblings are not related genetically, these results suggest that genetic differences account for about half of the differences between children in how much they watch television. This was mind-boggling because here was an archetypal measure of the environment showing as much genetic influence as we find for psychological traits.

I knew it was going to be difficult to convince psychologists that genetic differences influence television viewing because it was at that time a favourite 'environmental' measure. More data would help make the finding more convincing. During the home visits we also asked parents how much television they themselves watched, which meant that I could look at parent–child similarity. Despite the strong sibling results, I was not expecting much from these analyses because the reasons for watching television seem likely to be different for parents and children, which might mean that there is little resemblance between parents and their children. But even these parent–child results suggested substantial genetic influence. Non-adoptive parents and their children were significantly more similar (0.30) in how much television they watched than were adoptive parents and their adopted children (0.15).

The most astonishing result was that birth mothers' television viewing correlated significantly (0.15) with their adopted-away children's television viewing, even though these birth mothers had not seen their adopted-away children after the first week of life. This pattern of correlations for parents and their children indicates that about a third of the differences between children in their television viewing can be accounted for by genetic factors in their parents.

Although the results were consistent and strong, as I started to talk about these findings, some colleagues thought that this study might be a professional suicide note, because it was just too weird. This reaction made me hesitate to write a paper about it. At least I had by that time risen through the ranks to full professor with tenure, which provides a real sense of academic freedom to tackle unpopular topics. In the end, I decided that it would be a good opportunity to get psychologists' attention by showing that even an 'obvious' environmental measure like television viewing can show genetic influence.

Finally, in 1989, I wrote a paper about these findings. The paper's title was 'Individual Differences in Television Viewing in Early Childhood: Nature as Well as Nurture.' I tried to anticipate misunderstandings. I peppered the paper with phrases like, 'There can be no genes for television viewing, just as there are no genes for performance on IQ tests or for height' and 'Complex traits such as these are heritable but not inherited.'

After a protracted review process, the paper was published in 1990 in the first volume of the flagship journal of the new American Psychological Society. The reaction was not as bad as I had feared. Its reception was helped by a positive news story about the paper in the top science journal, *Science*, which does not often pay attention to psychological research. The story in *Science* ended by saying that 'the study is noteworthy because it adds TV viewing to the list of influences that are commonly viewed by psychologists as environmental, but which in fact are also partly genetic.'

Nonetheless, my television study has been used by critics of behavioural genetics as a poster child for how ridiculous behavioural genetic findings are. I happily ignore anti-genetics types who won't countenance the possibility of genetic influence, but I was bothered by an eminent behavioural geneticist who wrote, in a prominent review of behavioural genetic research: Genetic influence on 'TV viewing habits may be true ... but genetic analyses of such phenotypes are of uncertain meaning ... For example, no gene for TV watching, a behavioral phenotype non-existent three generations ago, could plausibly exist.'

Where to begin in responding to such comments? Who said anything about a 'gene for TV watching'? Why is a genetic analysis of individual differences in how much children watch television 'of uncertain meaning'? Television viewing has been used in thousands of studies as a measure of the environment, without anyone questioning its meaning. If the assumption that television viewing is an environmental measure were correct, our analyses should have found no genetic influence. Instead, our research showed that this 'environmental' measure is strongly influenced by genetic differences.

Another reason why this finding is held up for ridicule is that whether or not we watch television seems to be completely a matter of free will. We can click the television on or off as we please, so how can genes affect it? The answer is that free will is irrelevant in terms of genetic effects on complex traits. Genetics is about the extent to which inherited DNA differences account for differences between people. In other words, we can turn the television on or off as we please, but turning it off or leaving it on pleases individuals differently, in part due to genetic factors. Genetics is not a puppeteer pulling our strings. Genetic influences are probabilistic propensities, not predetermined programming.

What about the type of television children watch? The most reliable measure of television viewing that we had in the Colorado Adoption Project was overall viewing time, but we also had information on broad categories of programmes, such as comedies, drama and sport. I was amused to find that genetic influence was strongest for time spent watching comedies because I don't find most comedies funny. We didn't include this result in the paper because it was not statistically significant and I thought the paper was pushing the limits of weirdness without going into this.

By 1991 there were eighteen similar studies that reported results for genetic analyses of various environmental measures. I was amazed at how consistently these studies showed genetic influence. The average heritability was 25 per cent for these environmental measures. This is only half the heritability of most psychological measures but these are measures that are called 'environmental' because they were assumed to be purely environmental and yet a quarter of their variance was genetic in origin. To put this in perspective, accounting for 25 per cent of the variance of these measures with inherited DNA differences is off the scale of effect sizes in psychology, where we rarely explain more than 5 per cent of the variance. Also, this heritability of 25 per cent is the average across some measures that are more highly heritable, like controllable life events and children's television viewing, and those that are hardly heritable at all, like uncontrollable life events such as the death of a family member.

In 1991 I published a paper reviewing the results of these eighteen studies which I called 'The Nature of Nurture'. As a sign of the novelty of this finding, the paper was published with thirty-two commentaries by other researchers. Most commentaries were hostile or disbelieving.

This paper showed that heritability is not just limited to self-report questionnaires like life events, which involve perceptions. Genetic influence is just as strong for observational studies of parent–child interactions in which researchers rated specific behaviours of parents and children. Finding that genetic influence was just as substantial for objective observational measures as for subjective self-report measures suggests that genetic influence on experience is not just in the eye of the beholder. Genetic effects can be seen in actual behavioural interactions between parents and children.

Since then, more than 150 papers have looked at environmental measures in genetically sensitive studies. They consistently find substantial genetic influence and the average heritability is still about 25 per cent. What's new is that these studies have greatly extended the list of environmental measures that show genetic influence. For example, evidence for genetic influence has been found for home environments such as chaotic family environments, for classroom environments such as supportive teachers, peer characteristics such as being bullied, neighbourhood safety, being exposed to drugs, work environments and the quality of one's marriage. Results showing genetic influence are not limited to the classic twin design. They also emerged from studies of twins reared apart, other adoption designs and, most recently, from DNA studies.

Characteristics of adolescents' peer groups are especially highly heritable, such as the peer group's academic orientation or their delinquency. The reason for this high heritability may be that you can choose your friends but you cannot choose your family, as Harper Lee wrote in *To Kill a Mockingbird*. You passively share genes with your parents and siblings, which leads to correlations between genes and your family experiences. With friends, you can select individuals similar to you genetically, actively creating correlations between your genes and your experiences with friends.

Social support is another workhorse in psychological research on the environment. As we grow up and move into the world outside the family, our social networks grow to include adult friendships, co-workers, neighbours and, increasingly, social-media contacts. Support comes from these relationships in many forms, including financial and informational support, but in psychology social support usually refers to emotional support from relationships, a sense of belonging as well as warmth. Social support has been linked to mental and physical health and is an especially important ingredient in successful ageing.

As with other 'environmental' measures, no one had asked about possible genetic influence on individual differences in social support. It was assumed that social support predicts mental and physical health and successful ageing for environmental reasons. In the 1980s the opportunity to put this assumption to the test came from our SATSA study of twins reared apart and twins reared together. We included a traditional measure of social support that asks questions such as whether the interviewee had people who would help them if they were in trouble, who could drop in anytime and with whom they could share their innermost feelings. For each question, you are asked about the number of people who fill that bill and also about how satisfied you are with the level of support you perceive. The responses can be condensed into two factors: quantity, which is the size of the support network, and quality, which refers to satisfaction with the level of support. These two scales are only modestly correlated, which means that some people can be satisfied with a small network of support and some are not satisfied even though they have a large network.

For quality of support, we found that a third of the differences between people could be explained by genetic factors, but quantity of support showed no significant genetic influence. Why would quality of support show genetic influence but not quantity of support? In our paper describing these results we suggested that the answer might be that quality seems more subjective than quantity. More subjective measures catch genetic influence as perceptions filter through people's personality, memories and motivation. That's just a guess, though, and we still don't know why quality of support is more heritable than quantity of support. And this might be different now, with the prominence of social media, which seems more a matter of quantity than quality. A recent twin analysis showed that individual differences in the use of Facebook in young adults yielded a heritability of 25 per cent, although quantity and quality of social support were not distinguished.

Despite the initial disbelief and hostility to the early studies showing genetic influence on diverse 'environmental' measures, now, nearly thirty years later, the nature of nurture is widely accepted. Nonetheless, if Table 1 had included measures of the environment such as life experiences and social support, few people would have rated them as heritable.

Experience is not just something that happens to us. With all our genetically rich differences in personality, we differ in our propensity to experience life events and social support, to watch television and to get divorced.

Try to think of something in the psychological environment that cannot have anything to do with you and your genetics. Take weather, for example, the archetypal environmental factor over which we have no control. As Mark Twain supposedly quipped, 'Everybody talks about the weather, but nobody does anything about it.'

Can you do anything about the weather? Asked this way, the question sounds like an item on a psychotic-experiences questionnaire. Of course you can't change the weather. It is more useful to phrase the question in the language of individual differences, which is the bailiwick of behavioural genetics. Why do some people live in warm and sunny climes and others tolerate cold, wet places? One answer is that, although we cannot control the weather, we can control where we live. If you love being outside, or if you have seasonal affective disorder, you can consider moving to a climate that suits you. Being outdoorsy or being prone to depression is influenced in part by genetic factors. Moving to a climate that suits you is one way in which genetic differences could contribute to individual differences in responses to straightforward questions about the weather such as 'How often does the sun shine where you live?' You might live in a sunny place because you chose to live there.

Could evolutionary adaptation contribute to the heritability of one's climate? People whose ancestors have lived for many generations in a particular climate may have adapted evolutionarily. Certainly there are genetic adaptations for extreme climates. For example, the shorter limbs and squatter bodies of Eskimos may be an adaptation that allows them to conserve heat. Physical and physiological adaptations may also have evolved to accommodate life in the desert or at extreme altitudes. However, evolutionary adaptations such as these are about average differences between groups, whereas heritability is about individual differences. Twins, for example, grow up in the same group, so genetic causes of average differences between groups are not reflected in differences within pairs of twins. In the extreme, highly adaptive characteristics like bipedalism and frontal vision do not allow genetic variation, so that heritability would be 0. So, evolutionary adaptations for different groups are not likely to contribute to genetic differences between individuals within these groups.

A more likely source of genetic influence on weather is perceptions. I am an incorrigible optimist, seldom removing my rose-tinted glasses. Even though I live in England, which is not known for its constant sunshine, I find that when I look back at last summer's weather I recall that it wasn't too bad, remembering the sunny days spent sailing and swimming. I am always taken aback when others talk about last summer's rotten weather.

Some people say that these are just perceptions of the weather, not the real weather. In response, I would say that the psychologically effective environment is the perceived environment. That is, what we perceive about the environment is what we actually experience. Even if last summer's weather records show that it was the coolest and cloudiest summer in a decade, what matters to me is my memory of warm, sunny days. These perceptions can pick up genetic influence as they filter through my cognitive biases and personality. Although objective measures of the environment are useful, we should not discount the importance of subjective perceptions.

Once you start thinking about how much DNA matters, it is difficult to point to any psychological experiences completely devoid of possible genetic influence. For example, accidents are not always accidental. Some children have more accidents than others; the number of children's scrapes and bruises shows genetic influence. For adults, automobile accidents are not always accidental either, of course. Automobile crashes are often caused by reckless driving – driving too fast, taking chances or driving under the influence of alcohol and other drugs. Sometimes accidents do just happen, but genetic differences in personality can increase the likelihood of accidents happening.

The only events free from genetic influence are those over which we have little control, such as the death or illness of relatives and friends. As expected, research finds little genetic influence for these uncontrollable events. Nonetheless, our reaction to such events – our psychological experience of the events – can be influenced by our genetic make-up.

The importance of measures of the environment lies in their psychological impact. If genes affect environmental measures as well as psychological measures, this raises the possibility that genes also contribute to correlations between them. For example, good parenting correlates with children's good development, bad peers correlate with bad outcomes for adolescents, and stressful life events correlate with depression in adults. It was assumed that these correlations are caused environmentally. No one considered the possibility that genetics could also contribute to these correlations.

How can you tell if genetics contributes to these correlations? For the association between parenting and children, the most direct analysis is provided by the social experiment of adoption. Does parenting relate to children's outcomes as much in adoptive families, where parents and children share only nurture, as compared to non-adoptive families, where both nurture and nature are shared?

My interest in the nature of nurture began in the early 1980s, when I looked at early results from the Colorado Adoption Project, which included several measures of parenting. One was an observational measure of the home environment, which had recently been developed and is still the most widely used observational measure of the home environment of young children. It has the nice acronym HOME, which stands for Home Observation for Measurement of Environment. HOME includes forty-five items to record parents' specific behaviours towards the child, rather than general ratings. For warmth, for example, HOME includes items about caressing, kissing and talking to the child. Control was assessed with items like interfering with the child's actions and punishment. We assessed HOME when the children were aged one, two, three and four years old.

Collecting these HOME data during 2,000 visits to homes all over Colorado was a major investment of time and money. In 1984, when the visits were completed at the ages of one and two, I eagerly looked at the relationship between HOME and children's cognitive and language development. From the results of many other studies of non-adoptive families, I expected HOME to correlate about 0.5 with children's mental development and language development at the age of two. With relief, I saw that our data yielded these expected results for non-adoptive families, with correlations of about 0.5 between HOME and cognitive and language development. But when I looked at the correlations in the adoptive families they were significantly lower, only half the size of those in the non-adoptive families.

Because non-adoptive parents are genetically related to their children but adoptive parents are not, these results suggest that genes contribute to the correlation between HOME and children's cognitive development. We showed that about half of this correlation can be attributed to genetics.

These results mean that genetics is a 'third factor' that contributes to the correlation between parenting as assessed by HOME and children's cognitive development. That is, the correlation is not just due to HOME boosting children's cognitive development directly, nor is it just due to parents responding to differences in their children's cognitive ability. These two processes explain the correlation in adoptive families. The reason why the correlation is doubled in non-adoptive homes is that parents and offspring are related genetically.

How does genetics work as a 'third factor'? How is it possible that genes shared by parents and their children lead to correlations between such different things as parenting assessed by HOME and children's cognitive development? The key is to break out from the bonds of labels. The 'E' of HOME is 'environment' but what it assesses is parental behaviour. It is much easier to think how parental behaviour can be genetically correlated with children's behaviour. For example, parents with high scores on HOME are people who support and stimulate their children and are responsive to their needs. Suppose that these are brighter parents. Rephrasing the correlation between HOME and children's cognitive development as 'brighter parents have brighter children' makes the possibility of genetics as a 'third factor' seem plausible and probable.

In the Colorado Adoption Project we looked at hundreds of correlations for dozens of measures of parenting as they relate to dozens of measures of children's development. In our 1985 paper we concluded that genetics is generally responsible for about half of the correlation between parenting and children's psychological development.

Adoption studies like the Colorado Adoption Project are especially powerful for investigating the effects of family environments such as parenting on children's development. For measures of the environment outside the family, for instance life events in adulthood, a more general approach is *multivariate genetic analysis*. This type of analysis estimates genetic influence on the correlation between two traits rather than on the variance of each trait analysed separately.

Another 'aha' moment was when I realized that this general multivariate genetic approach to the analysis of two traits could also be used in twin studies to explore the role of genetics in the correlation between environmental and psychological variables. In the first study using this approach, in 1991, we looked at the correlation between social support and well-being in the Swedish study of middle-aged twins reared together and reared apart. Social support correlates about 0.25 with well-being, a correlation that, as usual, had been interpreted environmentally: Social support causes well-being. To the contrary, we found that genetics accounts for over half of the correlation.

Since 1991 more than a hundred studies of this sort have been reported, and they keep on coming. I have tried to review these studies, but I gave up, for two reasons. One reason is that the field is growing faster than I can assimilate it. The more important reason is that most studies report a genetic analysis of the correlation between a single environmental measure and a single psychological outcome. This is a problem because there are many environmental measures and many psychological measures so there are countless combinations of the two. This leads to a sprawling literature with few attempts to replicate specific results, which scuppers attempts to review them systematically.

Despite the difficulties in summarizing these studies systematically, they tell a simple story. This story is the same as that told by the original 1985 CAP paper and the 1990 SATSA paper: Genetics typically accounts for about half of the correlation between environmental measures and psychological traits. This finding about the nature of nurture is one of the most unexpected and important examples of how DNA makes us who we are. Instead of assuming that correlations between the 'environment' and psychological traits are caused environmentally, it is safer to assume that half of the correlation is due to genetic differences between people. This research is also important because it shows how we can study 'true' environmental effects controlling for genetics. This will be a major direction for research as the DNA revolution takes hold.

The nature of nurture suggests a new way of thinking about experience. In the past, psychologists assumed that the environment is what happens to us passively, but genetic research on the nature of nurture suggests a more active model of experience. Psychological environments are not 'out there', imposed on us passively. They are 'in here', experienced by us as we actively perceive, interpret, select, modify and even create environments correlated with our genetic propensities. Our genetically rich differences in personality, psychopathology and cognitive ability make us experience life differently. For example, genetic differences in children's aptitudes and appetites affect the extent to which they take advantage of educational opportunities. Genetic differences in our vulnerability to depression affect the extent to which we interpret experiences positively or negatively. This is a general model for thinking about how we use the environment to get what our DNA blueprint whispers that it wants. This is the essence of the nature of nurture.

4

DNA matters more as time goes by

As you go through life, would you imagine the effects of heredity become more important or less important? Most people will usually guess 'less important', for two reasons. First, it seems obvious that we are continually and cumulatively buffeted by environmental winds. The longer we live, the more we experience the impact of parents, friends, relationships and jobs, as well as accidents and illnesses. Second, people mistakenly believe that genetic effects never change from the moment of conception – that we inherit our DNA from our mother and father, and that it doesn't alter from the moment egg meets sperm.

From this perspective, one of the big findings from behavioural genetic research is counterintuitive: genetic influences become more important as we grow older. No psychological trait shows less genetic influence with age, but the domain where heritability increases most dramatically during development is cognitive ability.

There are many types of cognitive abilities – for example, verbal and spatial – but in fact you are more likely to have one if you have the other. People with higher ability for memory, say, tend to have higher ability for all the other forms of intelligence. People often think they are good at either literature or maths, for example, but in fact they are more likely to be good at both if they are naturally skilled in one, although there are exceptions.

The construct of intelligence captures what diverse cognitive tests have in common, which is why intelligence is often referred to as *general cognitive* ability, or *g*. 'Intelligence tests' usually include a dozen verbal and non-verbal tests and summarize performance as a total score called an IQ score, which is an acronym for an outdated concept, the 'intelligence quotient'.

According to the majority view of intelligence researchers, the core of intelligence is 'the ability to reason, plan, solve problems, think abstractly, comprehend complex ideas, learn quickly, and learn from experience'. Intelligence is important scientifically and socially. Scientifically, intelligence reflects how the brain works, not as specific modules that light up in brain-imaging studies, but as brain processes working in concert to solve problems. Socially, intelligence is one of the best predictors of educational achievement and occupational status.

During the past century, genetic research on intelligence was in the eye of the storm of the nature–nurture debate in the social sciences. The debate was driven by misplaced fears about biological determinism, eugenics and racism. This controversy raised the threshold for acceptance of the importance of genetics. Genetic research exceeded this threshold with bigger and better studies stockpiling evidence consistently showing that genetic differences between people account for about half of their differences in tests of intelligence. This general estimate of 50 per cent heritability masks an intriguing finding, which is how heritability changes over the course of our lives.

In 1983 I was part of an American delegation invited to go to the Soviet Union to study children's development in daycare centres, of which the Soviets were justifiably proud. The draw for us was to be able to go to parts of the Soviet Union that Westerners were rarely able to see in those days. I wondered why I had been invited because at that time my research was showing genetic influence in infancy and genetics was not politically correct in the Soviet Union because the environment was assumed to be all important. I came to see that the notion of genetics was in fact acceptable to the Soviets when it comes to young children because the rationale for their programme of intensive communal early childhood care was to acculturate children into communist society, erasing traces of their animal nature, which includes their genetic predispositions. So it was tolerable to demonstrate that we have genetic influence early because it was assumed that it could not be important in later development.

No evidence existed in support of this Soviet hypothesis that heritability disappears after childhood. Instead, research at that time was beginning to show the opposite: that DNA matters more as time goes by. The Louisville Twin Study first suggested that heritability increases for intelligence during infancy and childhood. In 1983 it reported results from a twenty-year study of 500 pairs of twins assessed fourteen times from infancy to adolescence. Identical twins became more similar for intelligence from infancy to adolescence, with identical twin correlations increasing from about 0.75 to 0.85. In contrast, fraternal twins became less similar, from about 0.65 to 0.55. Because heritability is estimated from the difference between identical and fraternal twin correlations, this pattern of results suggested increasing heritability – from about 20 per cent in infancy to about 60 per cent in adolescence.

Although the longitudinal results showed a consistent pattern of increasing heritability, the relatively small sample size of 500 pairs of twins did not have sufficient power to show that this change was statistically significant. Nonetheless, a dramatic confirmation of this finding came from our Colorado Adoption Project. Correlations between intelligence of non-adoptive parents and their children increased from about 0.1 in infancy to 0.2 in childhood to 0.3 in adolescence, as many other studies have shown. The most remarkable finding was that the same pattern of increasing resemblance was found for adopted children and their biological parents whom the adopted children had not seen since the first few days of life. By sixteen years of age, the correlation for intelligence was the same for adopted children and their biological parents as for children reared by their biological parents. The correlations between these adopted children and their adoptive parents, who share nurture but not nature, hovered near zero.

Further support for the hypothesis of increasing heritability came from a 2010 consortium of twin studies that brought together data on intelligence for 11,000 pairs of twins from four countries, a larger sample than all previous studies combined. These studies found that heritability of intelligence increased significantly from childhood to adolescence to young adulthood, from 40 to 55 to 65 per cent.

Finally, in 2013, a meta-analysis brought together results from all twin and adoption studies of intelligence and confirmed the developmental increase in heritability. These studies focused on development up to early adulthood because this is the age of most samples in behavioural genetic research. The few available studies of later life suggest that increasing heritability continues throughout adulthood to about 80 per cent heritability at the age of sixty-five.

The heritability of 50 per cent for intelligence is just the lifetime average across all studies. The impressive increase in heritability from 20 per cent in infancy to 40 per cent in childhood to 60 per cent in adulthood stands out from other traits that show little developmental change in heritability, most notably personality and school achievement.

In this context, results for school achievement are surprising. Because intelligence correlates substantially with school achievement, you would expect school achievement to show a similar pattern of increasing heritability. However, we find no developmental change in heritability for school achievement for any subjects in the longitudinal TEDS twin study, even though we find increasing heritability for intelligence. In fact, heritability of school achievement is about 60 per cent across the school years, higher than the heritability of intelligence, which is about 40 per cent.

How can this be? One possible explanation is that universal education in the early school years reduces environmental disparities in skills like reading and maths which are targeted by tests of school achievement, and this leads to high heritability even in the first few years of school. In contrast, schools do not teach intelligence, so its heritability increases during development as children select and create their own environments correlated with their genetic propensities for learning. In other words, teaching basic literacy and numeracy skills in the early school years largely erases environmental disparities, leaving genetics as the primary cause of differences between children in these skills. The heritability of intelligence increases during the school years so that, by secondary school, it catches up to the heritability of school achievement. Moreover, once children achieve basic literacy and numeracy skills, they can use these skills as tools for learning in general, which contributes to the genotype–environment correlational processes responsible for the increasing heritability of intelligence.

This may be a general explanation for the huge increase in the heritability of intelligence across development. Although our inherited DNA sequence does not change after the moment of conception, the effects of genes can change as time goes by. For example, male pattern baldness is highly heritable but the effects of these genes do not show up until hormones change in mid-life. An important psychological example is schizophrenia, for which the average age of onset is early adulthood. It is difficult to detect any differences in childhood for individuals who are later diagnosed as schizophrenic. It is likely that the genes that contribute to the disorganized thinking, hallucinations and paranoia characteristic of schizophrenia do not have their effect until the brain has developed to a high level of symbolic reasoning in early adulthood.

One possible explanation for the increasing heritability of intelligence is that more genes come online to affect intelligence, perhaps because the brain becomes increasingly complex. However, this reasonable hypothesis seems unlikely because genetic research across age shows that the same genes affect intelligence from childhood to adulthood. That is, genes are largely responsible for stability from age to age, whereas the environment is responsible for age-to-age change, which leaves open the question of why heritability increases.

This finding about genetic stability comes from studies called *longitudinal* studies measuring twins repeatedly over the years. Rather than estimating the genetic and environmental contributions to the variance of intelligence at one age, it is possible to estimate the genetic and environmental origins of age-to-age change and continuity. Using multivariate genetic analysis, mentioned earlier, we can study the extent to which genetic effects at one age correlate with genetic effects at another age, or genetic correlation. In essence, instead of correlating twins' scores at one age, multivariate genetic analysis correlates one twin's score at one age with the other twin's score at another age and compares these cross-age twin correlations for identical and fraternal twins.

This type of analysis shows that genetic effects on intelligence are highly stable from age to age. For example, in TEDS, genetic effects on intelligence in Year 2 correlate 0.7 with genetic effects on intelligence in Year 4. Genetic correlations from age to age are even greater after childhood. A recent DNA study strongly supports these results from twin studies, finding 90 per cent overlap in the genes that affect intelligence in childhood and adulthood.

If genetic effects are highly stable from age to age, how can the heritability of intelligence increase so much during development? The most plausible possibility is that slight nudges from genetics early in development are magnified as time goes by. That is, the same genetic factors snowball into larger and larger effects, a process that is known as *genetic amplification*.

Genetic effects could be amplified as we increasingly select, modify and create environments correlated with our genetic propensities. For example, children with a genetic propensity for high intelligence are likely to read books and select friends and hobbies that stimulate their cognitive development. This is the active model of experience mentioned earlier. Although twin studies support this model, the DNA revolution will provide definitive results. As we begin to find the DNA differences that account for the heritability of intelligence at each age, the amplification hypothesis predicts that the same DNA differences will be associated with intelligence in childhood, adolescence and adulthood, but they will have a bigger effect as time goes by.

I like the idea that we grow into our genes. The older we get, the more we become who we are genetically. To some extent, especially for cognitive ability, this means we become more like our parents as we age. Perhaps this is why people, as they get older, often seem to fear that they are becoming just like their parents.

5

Abnormal is normal

Fifty per cent of us will have a diagnosable psychological problem in our lifetime and 20 per cent will have had one within the last year. The cost in terms of suffering to patients and their friends and relatives, as well as the economic costs, make psychopathology one of the most pressing problems today. Although the problems are real, the issue that this chapter addresses is that psychological problems are diagnosed as if they are diseases that you either have or don't have. This either/or mindset means that scientists have tried to look for *the* cause of the disorder, something that makes 'us' different from 'them'. This view is deeply engrained in psychiatry, which follows the medical model of illness, treating mental disorder as if it were a physical disease like infection that has a simple, single cause.

Genetic research shows that the medical model is all wrong when it comes to psychological problems. What we call disorders are merely the extremes of the same genes that work throughout the normal distribution. That is, there are no genes 'for' any psychological disorder. Instead, we all have many of the DNA differences that are related to disorders. The salient question is how many of these we have. The genetic spectrum runs from a few to a lot, and the more we have, the more likely we are to have problems.

In other words, the genetic causes of what we call disorders are quantitatively, not qualitatively, different from the rest of the population. It's a matter of more or less (quantitative), not either/or (qualitative). This might seem like an arcane academic issue but this finding is completely changing clinical psychology and psychiatry, especially as the DNA revolution comes along. It means there are no

58

disorders – they are just the extremes of quantitative dimensions. That is what is meant by the slogan 'Abnormal is normal'.

This chapter begins this important tale as it unfolded and then explores its implications.

The first hint came from twin and adoption research that investigated links between diagnosed 'cases' and dimensional measures of relevant traits. For example, diagnosed reading disability can be compared to dimensional measures of reading ability that assess reading quantitatively from poor readers to good readers. Reading disability is a diagnosis of reading problems that is made to sound like a 'real' medical disorder by being given a Greek name, dyslexia. Medicalization of psychological problems is typical – for example, problems with learning arithmetic are given a diagnosis of dyscalculia, and attentional problems are called attention deficit hyperactivity disorder or hyperkinesis.

Genetic analyses investigating links between qualitative disorders and quantitative dimensions involve a type of multivariate analysis which examines the genetic links between traits, as mentioned earlier. In this case, multivariate genetic analysis looks at the genetic correlation between a categorical (qualitative) diagnosis and a continuous (quantitative) dimension. Using reading as an example, we correlate one twin's diagnosis (yes or no) with the co-twin's quantitative reading score, and compare these 'cross-correlations' for identical and fraternal twins. Multivariate genetic analyses of this type find strong genetic links between diagnoses and dimensions, meaning that the genes that contribute to the diagnosis are the same genes responsible for the dimension.

This research indicates that the same genes are responsible for reading disability and reading ability. Similar results have been found for other psychological disorders, suggesting that there are no genes *for* psychological disorders – they are the same genes responsible for heritability throughout the normal distribution, from those few people with very low genetic risk to the many people with average genetic risk to the few people with very high genetic risk.

Evidence of this sort indicates that what we call disorders are merely the quantitative extreme of the same genetic effects that operate throughout the distribution. In other words, we all have DNA differences associated with how well we read. How good or bad our reading is depends on how many of these DNA variants we inherit. From a genetic perspective, abnormal disorders are the extreme of normal dimensions. As we will see later, this new view of the abnormal as normal is changing everything in clinical psychology, from diagnosis to treatment.

Rather than describing this complex type of twin analysis in detail, it is easier to see why the abnormal is normal if we jump ahead to the DNA revolution. As detailed in Chapter 10, a DNA difference in a gene called *FTO* is more frequent in cases of obesity than in control groups. But it is not a gene 'for' obesity. The DNA difference is associated with a six-pound increase in body weight for thin people as much as for heavy people. That is, if you have this DNA difference but your sibling does not, you are likely to weigh more than your sibling. This is the case whatever size you and your sibling are.

This sort of finding has emerged time and time again in other DNA research on disorders. Genes originally identified because they are associated with a common disorder turn out to be associated with normal variation throughout the distribution. There is a continuum of genetic influence from one extreme to the other. In other words, as we find genes associated with reading disability, these DNA differences will not be 'for' reading disability. They will be related to the entire distribution of reading ability. These DNA differences will make good readers read slightly less well than other good readers without these genetic variants. Conversely, as we find genes associated with reading ability, the same genes will predict reading problems.

When we talk about genetics, it is easy to slip into thinking about the gene for this and the gene for that. I call this the *OGOD hypothesis*, for 'one gene, one disorder', which is misleading. Our species has thousands of single-gene disorders, but they are rare. In contrast, common disorders, including all psychological disorders, are not caused by a single gene.

A single-gene disorder means that a single mutation is necessary and sufficient for the disorder. For example, Huntington's disease is a single-gene disorder that damages certain nerve cells in the brain. It develops in adulthood and gets progressively worse over time, after twenty years leading to complete loss of motor control and intellectual function. The DNA variant is 'necessary' because you only get Huntington's disease if you have the mutation for Huntington's disease. It is 'sufficient' because, if you inherit the mutation for Huntington's disease, you will succumb to the disease.

For a hard-wired single-gene disorder like Huntington's disease, the genetic effect is qualitative, not quantitative. In this case, you can talk about a gene 'for' the disorder. But even though there are thousands of single-gene disorders, they are all rare. No single-gene causes of common psychological disorders have been found.

The genetic architecture of psychological disorders is the opposite of the OGOD hypothesis. The high heritability of psychological disorders is caused by many DNA differences, each with small effects. None of these DNA differences are necessary or sufficient to develop a disorder. Finding many such small genetic effects means that they must be distributed quantitatively in a normal, bell-shaped curve. For a particular disorder, depression for example, suppose 1,000 DNA differences are found between cases of depression and non-depressed control groups. These DNA differences are not exclusive to people diagnosed with depression. In the population, the average person may have 500 of these 1,000 depression-causing DNA differences. These people will have an average genetic risk for depression. Some people with few of these DNA differences will have lower than average risk for depression. And people with more than the average number of these DNA differences are more likely to be depressed.

This is exactly the way genetic influence works for all common disorders. Later, we will consider *polygenic scores* comprised of thousands of DNA differences identified by their association with psychological disorders. The point here is that these polygenic scores are always perfectly normally distributed, meaning that they predict variation throughout the distribution – from people who are hardly ever depressed to those who sometimes get depressed to people who are chronically depressed. These polygenic scores predict whether someone is diagnosed as depressed or not only because these people are at the extreme of the normal distribution of genetic risk. The abnormal is normal in the sense that we all have many of the DNA differences that contribute to the heritability of any psychological disorder. Whether or not we reach some arbitrary diagnostic cut-off depends on how many of these DNA differences we have.

This genetic research leads to a momentous conclusion: There are no qualitative disorders, only quantitative dimensions. Psychological problems like depression, alcohol dependence and reading disability are serious. The more extreme the problem, the more likely it is to affect the individual, their family and society. But because the genetic risk is continuous, it makes no sense to try to reach a decision about whether someone 'has' the disorder or not. There is no disorder – just the extremes of quantitative dimensions. People differ in how depressed they are, how much alcohol they consume and how well they read, but these problems are part of the normal distribution. A shift in vocabulary is needed so that we talk about 'dimensions' rather than 'disorders'.

Another important implication of the abnormal-is-normal finding is that we cannot cure a disorder because there is no disorder. Success in treatment should be viewed quantitatively, as the degree to which a problem is alleviated. We will return to these issues in the last chapter because the DNA revolution will bring these issues to life – to all of our lives.

This view of what we call abnormal as part of the normal distribution of differences is already changing the way we think about mental health and illness. In the most recent diagnostic manual of psychopathology, this trend is reflected in renaming some disorders as spectra, which is another word for dimensions. Schizophrenia is now schizophrenia spectrum disorder; autism is autistic spectrum disorder. This is why people now refer to someone as being 'on the spectrum', regardless of whether they actually are. This is a nod towards a quantitative dimensional approach.

The normal-is-abnormal view is much more radical. We are not just conceding a bit of grey space in between normal behaviour and diagnosed disorders like schizophrenia and autism, setting up yet another diagnostic category called 'spectrum disorder'. We are saying that the distinction between normal and abnormal is artificial. The abnormal is normal.

Because the notion of abnormal versus normal is so deeply

engrained and so difficult to escape, another example is warranted. This one is facetious but it gets to the heart of the matter. Imagine we discover a new disorder, giantism. This disorder, which we will diagnose on the basis of height greater than 196 cm (6 feet 5 inches), has a frequency of 1 per cent. DNA differences found to be associated with giantism will also be associated with individual differences in height throughout the distribution – for short people as well as tall ones. The point is that height and its genetic basis are perfectly normally distributed. There is no abnormal, just the normal distribution with its normal extremes. It won't help to create another diagnostic category of 'almost a giant'.

Why would we create a disorder of giantism when height is so clearly a continuous trait? It doesn't make sense. I would argue that it is just as nonsensical to create distinct disorders for any problems – physical, physiological or psychological. They are merely the quantitative extremes of continuous traits.

For psychological problems like reading disorders and depression, it is easy to see how children are more or less reading disabled or enabled and how adults are more or less depressed. But when you get to rarer disorders like schizophrenia and autism, it is tempting to fall back into the either/or mindset. The behavioural symptoms used to diagnose schizophrenia and autism are so severe that it seems implausible to say that individuals with these disorders are merely the extreme of the normal distribution. In other words, how can you be just a little schizophrenic or just a little autistic? Although individuals institutionalized with a diagnosis of schizophrenia exhibit bizarre behaviour, schizophrenia includes symptoms such as disorganized thoughts, dissociation and unusual beliefs as well as more severe symptoms like hallucinations and delusions. Who has not sometimes experienced some of these symptoms? Whether we are diagnosed as schizophrenic has to do with how severe our symptoms are and how much they affect our lives and the lives of others.

Perhaps there is a threshold at which risk tips an individual over the edge to become 'really' schizophrenic or autistic. Risks could be quantitative but the result could nonetheless be qualitative in the sense that people who fall over that edge are 'really' schizophrenic or autistic. Coming close to the edge doesn't count. Physiological disorders like heart attacks and strokes are held up as examples of this cliff edge. Lots of things contribute to your risk but you either have a heart attack or you don't. But this is not true. Heart attacks and strokes are often so mild we don't know that we have had one. Even these extreme examples of physiological disorders are a matter of more or less, not either/or. This is also the case for disorders like schizophrenia and autism – there is no threshold that a person crosses where they tumble down into 'real' schizophrenia or autism.

For some physiological problems it is easy to assess the dimension underlying the disorder, for example, blood pressure is the dimension that underlies hypertension – indeed, it is how hypertension is diagnosed. Similarly, for some psychological problems dimensions that underlie disorders seem obvious. For example, tests of reading ability are used to diagnose reading disability. Similarly, problems of hyperactivity can be assessed as a dimension from little to lots of activity. Depressive disorder is at the extreme of a dimension of mood. Although some problems such as schizophrenia and autism have symptoms so severe as to seem outside the normal distribution, if we accept that we all have thought disorders to some extent sometimes, we can assess these symptoms quantitatively, if we stop being obsessed with diagnosing whether people 'have' the disorder or not. In the same way, we can assess autistic symptoms such as problems with social relationships and communication quantitatively.

One issue that comes up in thinking about the relationship between dimensions and disorders is identifying the 'other end' of the distribution of problem behaviour. For example, with reading disability, it seems obvious that the other end of the distribution involves good reading. But it's not so simple. Does the other end of the distribution involve being good at basic reading processes like decoding and fluency, or being good at higher-level processes like comprehension? Or does it involve all these components of reading? Is happiness the other end of the dimension of depression? What is the other end of the dimension for poor attention? Is it simply being very attentive or could the other end involve different kinds of problems, like compulsiveness?

As we will see later, the DNA revolution will put this issue front and centre in clinical psychology and psychiatry. The polygenic scores that predict genetic liability for 'disorders' are perfectly normally distributed. Therefore, we can, for the first time, investigate individuals at the 'other end' of the normal distribution of polygenic scores to find out who they are.

The most general implication of this view of the abnormal as normal is that there is no 'us' versus 'them'. We all have DNA differences that affect our risk for psychological problems. The more of these DNA differences we have, the more problems we are likely to have. It's all quantitative – a matter of more or less.

6

Generalist genes

Until now, psychologists have had to rely on behavioural symptoms to diagnose disorders. For example, hallucinations, delusions and paranoia are signs of schizophrenia. Severe swings in mood signal bipolar disorder. Poor attention span and high activity levels suggest attention deficit hyperactivity disorder. Although these are all important behavioural problems, the way they are lumped together in current diagnostic classification schemes is not supported by genetic research. For the first time, genetics offers a causal basis for predicting disorders rather than waiting until symptoms appear and then trying to use these symptoms, rather than causes, to diagnose disorders. Studies of genetic causes chart a map of disorders that is almost unrecognizable from our current symptom-based diagnoses of disorders. That is, instead of finding distinct genetic influences that correspond to diagnoses, genetic effects are splashed out across many disorders. Genetic effects tend to be general rather than specific, which is why I call this topic *generalist genes*.

Family studies first suggested that genetic effects might be general across disorders rather than specific to each disorder. These disorders do not 'breed true' – parental psychopathology predicts that the children of such parents are more likely to have psychological problems, but not the same disorder as the parent. For example, a parent might receive a diagnosis of depression and their offspring a diagnosis of antisocial behaviour. Developmental studies also show that one disorder often morphs into another.

Since the 1990s twin studies have also hinted at generalist genes in multivariate genetic analyses that investigated the genetic links between pairs of disorders. One of the first hints came from research that showed that generalized anxiety disorder and major depressive disorder are the same thing genetically. Inherited DNA differences contribute substantially to your risk of being anxious or depressed but they do not specify whether you will be diagnosed as anxious or depressed. Whether you become anxious or you become depressed is caused by environmental factors. In other words, genetic risks are general across disorders; environmental risks are specific to a disorder. Generalist genes are not limited to cases diagnosed with disorders. The same result emerged from two dozen twin studies that looked at the genetic overlap between dimensions of anxiety symptoms and dimensions of depression symptoms.

Hundreds of studies later, the genetic architecture of psychopathology suggests just three broad genetic clusters, in contrast to the dozens of disorders in psychologists' diagnostic manuals. One cluster includes problems like anxiety and depression, which are called *internalizing* problems because they are directed inward. The second genetic cluster, *externalizing* problems, includes problems in conduct and aggressiveness in childhood, and, in adulthood, antisocial behaviour, alcohol dependence and other substance abuse. Psychotic experiences such as hallucinations and other extreme thought disorders form the third genetic cluster, which includes schizophrenia, bipolar disorder and major depression.

Within these three genetic clusters genetic correlations are typically greater than 0.5, meaning that if you found a DNA difference associated with one type of problem, there is a fifty-fifty chance that it would also be associated with other types of problems. Not all genetic effects are general – some genetic effects are specific to one disorder – but the surprise has been to find how general genetic effects tend to be. Recently it has been suggested that these three clusters also overlap to create a general genetic factor of psychopathology.

Most severe mental illnesses, or psychoses, show the effects of generalist genes. The first branching point in diagnosing psychoses separates schizophrenia and depressive disorders. This dividing point is so enshrined in diagnoses that the two diagnoses were until recently viewed as mutually exclusive. That is, if you were diagnosed as schizophrenic, you could not be diagnosed with bipolar disorder, a severe form of depression that alternates with mania. For this reason, it was a complete surprise to find that most DNA differences found to be associated with schizophrenia also showed associations with bipolar disorder, as well as with major depression and other disorders. Even though schizophrenia, bipolar disorder and major depressive disorder are the oldest and most consistently diagnosed disorders, they show the greatest genetic overlap. This means that we are going to have to tear up our diagnostic manuals based on symptoms.

Other DNA techniques described in later chapters are beginning to be used more generally to analyse genetic links between traits, and these studies confirm the important role of generalist genes in psychopathology first discovered in twin studies. The DNA revolution will lead to a fresh approach to psychopathology that focuses on genetically defined mental health and illness, not only for identification of problems but also for treatment and prevention, as discussed in the last chapter.

Generalist genes are not limited to the domain of psychopathology. Most genetic effects are also general across cognitive abilities. For example, cognitive abilities such as vocabulary, spatial ability and abstract reasoning yield genetic correlations greater than 0.5, even though these abilities are thought to involve very different neurocognitive processes. That is, when we find a DNA difference associated with one cognitive ability, there is a greater than 50 per cent chance that it will also be associated with other cognitive abilities. Some genetic effects are specific to each cognitive ability, but the surprise is that most genetic effects are general to all cognitive abilities.

This is why intelligence, more precisely called *general cognitive ability*, is such a powerful construct. It captures what is in common among diverse cognitive abilities. This makes intelligence a good target for hunting generalist genes.

Education-related skills such as reading, mathematics and science show even higher genetic correlations: about 0.7. One of my favourite examples of generalist genes involves reading. A test called the Phonics Screening Check was developed to distinguish two components of reading that were thought to be fundamentally different processes. One is the ability to read familiar words quickly and accurately (fluency). The other is the ability to sound out non-words (phonetics). A test like this is administered to all 600,000 five- and six-year-olds in the UK because there is an assumption that it separates out these two components of reading, fluency and phonetics.

The test involves reading aloud as quickly as possible a list of age-appropriate familiar words and 'non-words'. For example, familiar words might be 'dog' and 'exercise'. Non-words are word-like combinations of letters never seen before that are matched in difficulty level to the real words, such as 'pog' and 'tegwop'. The reasonable idea underlying this interesting test is that reading familiar words should be automatic, but non-words that children have never seen before require sounding them out, which is phonetics.

Reasonable ideas are often wrong, as in this case. The genetic correlation between reading familiar words and non-words is 0.9, making this one of the most powerful examples of generalist genes. That is, the same DNA differences are responsible for individual differences in fluency and phonetics, even though fluency and phonetics are thought to be completely different neurocognitive processes.

A recent example of the power of generalist genes comes from studies my team have done on spatial ability. We developed a dozen online tests with the goal of identifying specific components of spatial ability such as navigation, mechanical reasoning and the ability to visualize objects when they are rotated in two and three dimensions. Despite our best efforts to assess specific aspects of spatial ability, generalist genes overwhelmed specific genetic effects. Genetic correlations among the dozen spatial tests were on average greater than 0.8.

I find that a common reaction from psychologists to this evidence about generalist genes is disbelief. Some children with reading problems have no problem with mathematics, and vice versa. If genes are generalists, why do specific disabilities occur? First, there is less specificity than it might seem. Reading and mathematics performance correlate highly but, even so, on statistical grounds alone, some children are expected to be better in one area than the other because the correlation is not 1. Second, genes are also specialists – genetic correlations are not 1. It is not surprising that there are some specialist genes. The surprise is how general genetic effects are.

Generalist genes are also likely to be relevant to brain structure and function. Neuroscientists assume that different parts of the brain do specific things, an idea known as *modularity*. In contrast, generalist genes imply that individual differences in brain structure and function are largely caused by diffuse effects that affect many brain regions and functions.

The generalist-genes model makes more sense genetically and evolutionarily than the traditional modularity model. There are two great principles of genetics as they affect complex psychological traits like psychopathology and cognitive abilities as well as neurocognitive traits involving brain structure and function. First, genetic influence is caused by thousands of DNA differences of extremely small effect size; this is called *polygenicity*. Second, each DNA difference affects many traits; this is called *pleiotropy*. Given polygenicity and pleiotropy, it seems likely that generalist genes result in generalist brains.

It also makes sense to assume that the brain evolved as a general tool for solving problems. Natural selection did not care about making things easy for neuroscientists by creating neat modules with specific functions. In fact, the brain did not evolve, people did. Our ancestors' survival depended on how well their brainpower translated into behaviour. Individuals who were better able to solve problems, including life-and-death split-second decisions, were more likely to survive and reproduce. Individual differences in problem-solving scooped up whatever advantages they had wherever they occurred in the brain.

Generalist genes have not yet been investigated in neuroscience, in part because neuroscientists seldom consider individual differences. To study individual differences requires large sample sizes and this is difficult to do because brain-imaging studies are expensive. The DNA revolution will change this. I confidently predict that, come the DNA revolution, we will find that generalist genes are important at every step along the pathways from genes to brain to behaviour.

7

Why children in the same family are so different

Two of the five big findings from behavioural genetics are about the environment. The first is the nature of nurture, as discussed earlier. We stumbled over this phenomenon by noticing that what psychologists call 'environmental' measures often show genetic influence. This eventually led to a radical new view of how the environment works. The environment is not something 'out there' that happens to us passively. Instead, we actively perceive, interpret, select, modify and even create environments, in part on the basis of our genetic propensities.

The second big finding about the environment also began by bumping into something odd: Why are children who grow up in the same family so different? One sibling might be an extravert, the other withdrawn; one may be better at school than the other. We now know that genetics makes siblings 50 per cent similar, which means it also makes them 50 per cent different. But before genetics was taken seriously, it was a puzzle why children growing up in the same family, with the same parents, living in the same neighbourhood and going to the same school should be so different.

Siblings are not completely different, of course. For example, if one sibling is diagnosed as schizophrenic, their siblings have a 9 per cent risk of being schizophrenic, much greater than the rate of 1 per cent across the general population. Siblings correlate about 0.4 for intelligence. There was no problem explaining why siblings are similar. When psychology emerged as a science early in the twentieth century it was dominated by environmentalism, the view that we are what we learn. The family was the first and predominant source of how the environment makes us who we are. Belief in the power of the family environment made it easy to assume that nurture is the reason why

71

psychological traits run in families. Why are you similar to your siblings? Because you grew up in the same family.

Given the presumed power of the family environment, however, it was difficult to explain why siblings are so different. For example, more than 90 per cent of the time, when one sibling is diagnosed as schizophrenic, the other is not. The average IQ score difference between siblings is thirteen IQ points, not that far off the average difference of seventeen IQ points for pairs of individuals selected at random from the population.

Many famously different siblings come to mind. There is Bill Clinton and his ne'er-do-well half-brother Roger, whom the US Secret Service gave the codename 'Headache' and who was sent to prison for drug dealing. In fiction, sibling differences are central to many plots, such as Tom Sawyer and his brother Sid (Mark Twain in his autobiography admits that the fictional pair closely parallels the differences between him and his real-life brother Henry). Biographies often describe differences between siblings. Everyone who has written about William James, who founded American psychology, or his brother Henry James, the novelist, emphasized how different they were. Henry, by his own admission, was unconfident, aloof, lacking William's easy gregariousness and savoir-faire and envying his effortless talents and capabilities.

The answer to the fundamental question of sibling difference led to a breathtaking, almost unbelievable, view of how the environment works. Yet this finding lay unnoticed in the background of the first century of behavioural genetic research, which focused on nature versus nurture. Twin and adoption methods were designed to tease apart nature and nurture in order to explain family resemblance. For nearly all psychological traits, the answer to the question of the origins of family resemblance is nature – things run in families primarily for genetic reasons. However, the same research provided the best available evidence for the importance of the environment, because heritabilities are usually about 50 per cent, which means that half of the differences between people are due to the environment, not to genetics.

It was not until the 1970s that behavioural geneticists began to realize what this meant. We resemble our parents and our siblings because we are similar to them genetically, not because we grow up in the same environment and experience the same opportunities or traumas. In other words, growing up in the same family with someone does not make you resemble them beyond your genetic similarity. The astonishing implication from this research is that we would be just as similar to our parents and our siblings even if we had been adopted apart at birth and reared in different families. As unbelievable as this might seem, as we shall see, adoption research shows that this is literally true.

But this finding packs an even bigger punch. The goal of behavioural genetic research is not to understand why siblings are similar or different. Behavioural genetic research uses twins and adoptees to understand what makes all people, including only children, different. What this research implies is that, far from the family being a monolithic determinant of who we are, environmental influences shared by family members do not make a difference. This is an astonishing conclusion because these are just the environmental influences that psychologists consider when they talk about nurture.

Although this conclusion might seem incredible, we have already seen some of the data that leads to this conclusion, but the data have been in the background of research on nature and nurture that we considered earlier, when we used weight as an example. If growing up with someone makes you similar to them, adoptive relatives should be just as similar as genetic relatives. To the contrary, we saw that the weight of adopted children does not correlate at all with the weight of their adoptive siblings or parents with whom they share the same family environment.

Even more surprising is the point alluded to earlier: Adopted individuals are just as similar in weight to their biological siblings and parents when they are adopted apart at birth and reared in different families as are relatives who share their family environment. This is even the case for identical twins separated at birth. They are nearly as similar in weight in adulthood as identical twins reared together from birth.

Twin studies come to the same conclusion that growing up in the same family does not make family members similar in weight, unless they share genes. Twin studies estimate heritability of weight as 80 per cent, even though all the genetic data together estimate heritability as 70 per cent. Identical twins correlate 0.8, which means that genetic similarity completely accounts for their similarity in weight. Fraternal twins correlate 0.4, which is exactly what would be expected if heritability is 80 per cent, because fraternal twins are only 50 per cent similar genetically.

Although weight is a good example because there is so much relevant data, adoption and twin studies reach the same conclusion for all personality traits and psychopathology. Heritabilities are typically 50 per cent, which completely explains the similarity of relatives. The environment accounts for the other 50 per cent, but there is no evidence for shared environmental effects of growing up in the same family.

The absence of evidence for shared environmental influence has been found not only for traditional personality traits like extraversion and neuroticism but also for traits that might seem especially susceptible to parental influence, such as altruism, caring and kindness. These traits are components of a factor that personality researchers call *agreeableness*. I had always assumed that these traits would show shared environmental influence and was pleased that in the first genetic study of agreeableness we found that shared environmental influence accounted for at least 20 per cent of the variance. Unfortunately, subsequent research has not confirmed this finding and I reluctantly admit that even agreeableness shows no influence of shared environment. Grit is another personality trait that has been thought to be due to shared environmental influence, but it also shows the same results as other personality traits: moderate heritability and no shared environmental influence. Nurture does not teach children to be kind or gritty.

Model-fitting analyses that put all the data together consistently find that experiences shared by family members have no effect on individual differences. Family members are similar for all psychological traits but for genetic reasons. Growing up with a sibling does not make you similar to them beyond the similarity due to genetics.

The environment is an important source of differences between people, but it's not the shared family environment that psychologists assumed was important. *Non-shared environment* is the name I gave to this mysterious type of environmental influence that makes children growing up in the same family different from one another. Like heritability, shared and non-shared environment are anonymous components of variance, bottom-line estimates of what makes us different, without specifying which specific factors are responsible.

Shared environment refers to any non-genetic factor that makes family members similar. Once you take heritability into account, there is no family resemblance left to explain, which means that shared environment is negligible. Non-shared environment refers to the rest of variance not explained by heritability or shared environment. Like heritability, estimates of shared and non-shared environment describe 'what is' in a particular population at a particular time. These estimates are limited to things that make a difference in that population. Rare events like abuse can make a huge difference for the abused individual but do not account for much variance in the population.

This finding about the importance of non-shared environment was ignored when it was first noted in 1976 in relation to personality. It was controversial when I first reviewed the genetic research pointing to this phenomenon in 1987, and again in 1998, when a popular book tackled the topic. But now the finding is so widely accepted, at least among behavioural geneticists, that attention has switched to finding any shared environmental influence at all. For instance, delinquency in adolescence shows some shared environmental influence, meaning that you might be more likely to get into bad behaviour if your sibling does, although even here most of the environmental influence is non-shared.

Intelligence appears to be a major exception to the rule that environmental factors that affect psychological traits are non-shared. The correlation for half a dozen older studies of adoptive siblings was 0.25, suggesting that a quarter of the variance in intelligence can be explained by shared environment. However, in 1978, a study of adoptive siblings reported a correlation of 0 for adoptive siblings' intelligence. Although this might have been a failure to replicate, the authors noted that the adoptive siblings they studied were between sixteen and twenty-two years old, while adoptive siblings in all previous studies were children. Could it be that the importance of shared environment for intelligence drops out by adolescence? Subsequent studies of older adoptive siblings have found similarly low correlations for intelligence. The most impressive evidence comes from a ten-year longitudinal follow-up study of adoptive siblings. At the average age of eight, the adoptive siblings correlated 0.25 for intelligence. Ten years later, the same adoptive siblings correlated 0.

These findings, supported by twin studies, suggest that shared environment affects intelligence during childhood when children are living at home. But as children reach adolescence and their worlds extend beyond the family, the impact of shared environment becomes negligible. In the long run, shared environment is not an important source of individual differences in intelligence. It is interesting that while the impact of shared environment declines during adolescence, heritability increases steadily from childhood through adulthood.

School achievement is another apparent exception to the rule. Tests of school achievement in all subjects from science to the humanities typically estimate that 20 per cent of the variance in performance can be explained by shared environment. Does the effect of shared environment on school achievement diminish after adolescence, as it does for intelligence? The first genetic research on educational achievement at university suggests this might be the case. Shared environment had no effect on performance in STEM subjects (science, technology, engineering and mathematics) and accounted for only 10 per cent of the variance on performance in humanities subjects. The only other exceptions from the hundreds of traits that have been investigated are some religious and political beliefs, for which shared environment accounts for about 20 per cent of the variance.

What are these mysterious non-shared environmental influences which are the main environmental reason why people differ? Any environmental factor can be analysed as a potential source of non-shared environment simply by asking whether it makes siblings different. For example, parents do not treat their children exactly the same. Environmental factors outside the family – such as school, peers and relationships – can contribute to the non-shared experiences of siblings. Even events shared by siblings could be a source of non-shared environment if the event is experienced differently by siblings. For example, if parents in one family divorce, this is an event that affects all the children, but those children can still experience and perceive it differently. It is often harder on one sibling than the other, or one might take it more personally. Unless an environmental factor makes children in the same family different, it cannot be important in development.

Despite the many possible candidates, progress in identifying specific sources of non-shared environmental effects has been slow. There are three steps towards identifying non-shared environmental influences. The first is to identify environmental factors that differ between siblings. Summarizing a huge literature, siblings living in the same family have very different experiences. It is almost as if siblings are living in different families, especially when it comes to their perceptions of how differently they are treated by their parents. Early research focused on parents and siblings. In retrospect, it seems odd that so much research looking for factors that make family members different would focus on the family. Looking outside the family – school, peers, friends, for example – would seem a better bet for finding factors that make siblings different.

The second step is to show that these environmental differences make a difference psychologically. That is, parents might treat their children differently, but does it make a difference in how the children turn out? Only a few candidates for non-shared environment make it to this second step. One example is that differences in parental negativity towards their children relate to the children's differences in terms of how likely they are to become depressed. That is, the sibling treated more negatively is more likely to become depressed. But why would parents treat one sibling more negatively than another sibling? This leads to the third step.

The third step takes on board the nature-of-nurture phenomenon. 'Environmental' measures show genetic influence, and genetics typically accounts for about half of the correlation between environmental measures and psychological traits. In other words, siblings might be treated differently because they differ genetically. For example, differences in how negative parents are towards their children might be an effect rather than a cause of a child's depression. Very few candidates for non-shared environment are left at this third step.

In the 1990s my colleagues David Reiss and Mavis Hetherington and I conducted a ten-year longitudinal study of 700 families with adolescent siblings called 'Non-shared Environment in Adolescent Development' (NEAD). The aim of NEAD was specifically to

identify non-shared environmental influences in a genetically sensitive design. NEAD controlled for genetic differences using a unique
design that included twins, full siblings, half siblings and adoptive
siblings. NEAD found several environmental measures that made it
to the second step of showing differences within siblings that related
to differences in their psychological outcomes. The example mentioned above was one of the strongest non-shared environmental
associations in NEAD: Differences in parental negativity towards
their children related to the children's differences in depression as
well as to differences in antisocial behaviour.

But hardly any NEAD findings made it to the third step. The
apparent associations between non-shared environment and psychological outcomes are largely due to genetic differences. The first
report of this phenomenon came from NEAD, showing that genetics
was largely responsible for the association between differences in
parental negativity towards their children and the children's differences in their likelihood of becoming depressed or engaging in
antisocial behaviour. In other words, parents' negativity was a
response to, rather than a cause of, their children's depression and
antisocial behaviour. It's as if the parents and their children are in a
downward spiral, with negative feedback loops between the parents'
negativity and their adolescents' unpleasant behaviour. Interestingly,
most of the non-shared associations that make it to this third step
involve the 'dark side' of development, such as negative parenting and
negative outcomes like depression and antisocial behaviour.

Identical twins provide an especially sharp scalpel to dissect non-
shared environment while controlling for the possibility of genetic
effects. Because identical twins are identical genetically, these siblings
differ only for reasons of non-shared environment. However, studies
of differences between members of identical twin pairs have found
few associations between identical twins' environmental differences
and their psychological differences.

In desperation, we conducted several studies of identical twins who
differed the most in certain traits, for example, in school achievement. We interviewed the twins and their parents to see if we could
generate hypotheses about environmental factors that made the identical twins different. We probed with general questions, such as 'You

and your twin agree that you differed in how well you did at school; what do you think led to the difference?', and more specific questions based on a previous questionnaire they had completed. We didn't find much. For example, for some identical twin pairs the twin who did better at school in a particular subject said they had a better teacher or were more interested in the subject and worked harder. The overall impression we got was that the twins and their parents didn't really know what environmental factors made them so different.

It seems likely that the influence of non-shared environment comes from many experiences that each have a small effect. There might be so many experiences of such small effect that they are essentially idiosyncratic, meaning that it comes down to chance. Sometimes chance is writ large, as in the case of major illnesses or accidents or war experiences that dramatically alter the course of an individual's development. More surprising are the often seemingly trivial chance events that launch lives in slightly different directions with cascading effects as time goes by.

It is fascinating how often biographies and autobiographies point to chance, such as childhood illness, as critical in explaining why siblings are so different. One of my favourite examples is the story Charles Darwin told in his autobiography about how he ended up on his five-year voyage on the *Beagle*, which led him to the theory of evolution. Darwin wrote that 'The voyage of the *Beagle* has been by far the most important event in my life, and has determined my whole career; yet it depended on so small a circumstance as my uncle offering to drive me thirty miles to Shrewsbury, which few uncles would have done, and on such a trifle as the shape of my nose.'

Darwin's comment about his nose refers to the quixotic captain of the *Beagle*, Captain Fitz-Roy, whom Darwin met for the first time in Shrewsbury in Shropshire. Fitz-Roy nearly rejected Darwin for the trip because Fitz-Roy was a believer in phrenology, which used the shape of the head to predict personality. The shape of Darwin's nose indicated to Fitz-Roy that Darwin would not possess sufficient energy and determination for the voyage. In one of his few jokes, Darwin wrote that during the voyage Fitz-Roy became convinced that 'my nose had spoken falsely'.

Francis Galton, the nineteenth-century founder of behavioural genetics (and cousin of Charles Darwin), suggested the importance of chance when he commented about 'the whimsical effects of chance in producing stable results'. Sounding like a fortune cookie, Galton went on to say that 'tangled strings variously twitched, soon get themselves into tight knots'. In other words, minor chance events can have knock-on effects across time.

These examples, and Galton's metaphor of tangled strings getting into tight knots, imply that these chance events have a lasting impact. But the situation is even more complicated. Genetic research shows that non-shared environmental influences are not only unsystematic, in the sense that they are mostly a matter of chance, they are also largely unstable, that is, inconsistent across time. The research that persuaded me about this involved longitudinal analyses of identical twin differences. Identical twin differences for psychological traits, which can only be due to non-shared environment, are not stable across time. That is, the twin who is happier today might be the unhappy twin tomorrow. Identical twin differences are a bit more stable for cognitive abilities and school achievement than for personality and psychopathology, but not much. No identical twin differences have been shown to be stable over several years, which would be necessary if non-shared environment had enduring effects. This means that the non-shared environmental factors that make identical twins different are not stable. They are like random noise.

In 1987 I wrote about this as the 'gloomy prospect' – the possibility that 'the salient environment might be unsystematic, idiosyncratic, or serendipitous events'. In other words, the key environmental influence making us who we are might be down to chance, unpredictable events. To this gloomy list, I would now add that their effects don't last. All of this makes these events extremely difficult to study.

Rather than accepting this gloomy prospect at the outset, it made more sense scientifically to look for possible systematic sources of non-shared environmental effects. However, after thirty years of searching unsuccessfully for systematic non-shared environmental influences, it's time to accept the gloomy prospect. Non-shared environmental influences are unsystematic, idiosyncratic, serendipitous events without lasting effects. The systematic, stable and long-lasting source of who we are is DNA.

8

The DNA blueprint

In 2010 Michael Gove, the recently appointed UK Secretary of State for Education, decided that UK schools should go back to teaching reading using phonics to sound out letters and words. At that time the national curriculum used the 'look and say' technique, in which children learn whole words on sight and are expected gradually to pick up the ability to recognize letter sounds. To make sure that teachers are following through on this curriculum change, all Year 1 pupils are tested on the Phonics Screening Check.

The Phonics Screening Check, mentioned earlier, involves reading aloud as quickly as possible a list of forty age-appropriate familiar words and non-words. For example, some easy words are 'dog', 'big' and 'hot' and more difficult words are 'project', 'frequent', 'exercise'. Non-words are word-like combinations of letters the child has never seen before that are matched to the real words in difficulty level. They also vary from easy ('pog', 'dat', 'bice') to difficult ('supken', 'tegwop', 'slinperk'). The reasonable idea behind this interesting test is that reading familiar words should be automatic, but non-words that the child has never seen before require sounding out, which is phonetics.

How well children performed on the Phonics Screening Check was assumed to be due to how well their teachers taught phonics. Schools are named and shamed if their students do not meet national standards. As usual in education, genetics was not even mentioned in the debate surrounding the phonics test. However, when we administered the test in the Twins Early Development Study, we found that it was among the most highly heritable traits ever reported at this age, with heritabilities of about 70 per cent. This means that the test is not measuring how well children are taught reading. Instead, it is a sensitive measure of genetically driven aptitudes for learning to read. Environmental factors shared by children going to the same school as well as growing up in the same family account for less than 20 per cent of the variance in children's performance on the phonics test.

Education is the field that has been slowest to absorb the messages from genetic research. In other fields, especially psychology, we have come a long way from the environmentalism that assumes that we are what we learn. Finding that heritability accounts for about half of the psychological differences between people means that genetics is by far the most important systematic influence on psychological outcomes. Genetics is the major reason why people differ in personality, mental health and illness, and learning and cognitive abilities. DNA is the blueprint for who we are.

The environment accounts for the other half of the variance but, as we saw in the previous chapter, it is not the environment as we had understood it that was important. We know that the environment makes a difference because heritability is only about 50 per cent, but the salient environmental influences are not the shared, systematic and stable effects psychologists had assumed were important in development. Moreover, research on the nature of nurture has demonstrated that what look like environmental effects are to a large extent really reflections of genetic differences.

Together, these findings point to a new view of human individuality that has sweeping implications for individuals and society. This chapter will explore these implications for parenting, schooling and life events, and the next chapter will consider their implications for equal opportunity and meritocracy.

PARENTS MATTER, BUT THEY DON'T MAKE A DIFFERENCE

Parents obviously matter tremendously in their children's lives. They provide the essential physical and psychological ingredients for children's development. But if genetics provides most of the systematic variance and environmental effects are unsystematic and unstable, this implies that parents don't make much of a difference in their children's outcomes beyond the genes they provide at conception. We saw in the previous chapter that shared environmental influence hardly affects personality, mental health or cognitive abilities after adolescence. This even includes personality traits that seem especially susceptible to parental influence such as altruism, kindness and conscientiousness. The only exception from hundreds of traits that shows some evidence of shared environmental influence is religious and political beliefs. As a parent, you can make a difference to your child's beliefs, but even here shared environmental influence accounts for only 20 per cent of the variance.

Furthermore, when differences in parenting correlate with differences in children's outcomes, the correlation is mostly caused by genetics. These correlations are caused by the nature of nurture rather than nurture. That is, parenting correlates with children's outcomes for three reasons considered earlier. One reason is that parents and their children are 50 per cent similar genetically. Put crudely, nice parents have nice children because they are all nice genetically. Another reason is that parenting is often a response to, rather than a cause of, children's genetic propensities. It is awkward to be an affectionate parent to a child who is not a cuddler. Finally, children make their own environments, regardless of their parents. That is, they select, modify and create environments correlated with their genetic propensities. Children who want to do something like play sports or a musical instrument will badger their parents to make it happen.

In essence, the most important thing that parents give to their child is their genes. Many parents will find this hard to accept. As a parent, you feel deep down that you *can* make a difference in how your children develop. You *can* help children with their reading and arithmetic. You *can* help a shy child overcome shyness. Also it seems as if you must be able to make a difference because you are bombarded with child-rearing books and the media telling you how to do it right and making you anxious about doing it wrong. (These books are, however, useful in providing parenting tips, for example, about how to get children to go to sleep, how to feed fussy children and how to handle issues of discipline.)

But when these best-selling parenting books promise to deliver developmental outcomes, they are peddling snake oil. Where is the evidence beyond anecdotes that children's success depends on parents being strict and demanding 'tigers' or giving their children grit? There is no evidence that these parenting practices make a difference in children's development, after controlling for genetics.

This conclusion is also difficult for many of us to accept in relation to our own parents. As you think about your childhood, your parents no doubt loom large, seeming to be the most significant influence in your life. For this reason, it is easy to attribute how we turned out, in good ways and bad, to our parents. If we are happy and confident, we might credit this to our parents' love and support. Or if we are psychologically damaged, we might blame this on inadequate parenting. However, the implications of genetic research are just as applicable here. These differences in parenting are not correlated with differences in children's outcomes once you control for genetics. Your parents' systematic influence on who you are lies with the genes they gave you.

If you are still finding it difficult to accept that parenting is less influential than you thought, it might be useful to review two general caveats about genetics that we considered earlier. The first caveat is that genetic research describes what is, not what could be. Parents *can* make a difference to their child but, on average in the population, parenting differences don't make a difference in children's outcomes beyond the genes they share. Parents differ in how much they guide their children in all aspects of development. They differ in how much they push their children's cognitive development, for example in language and reading. Parents also differ in how much they help or hinder their children's self-esteem, self-confidence and determination, as well as more traditional aspects of personality such as emotionality and sociability. But in the population, these parenting differences don't make much of a difference in their children's outcomes once genetics is taken into account. Over half of children's psychological differences are caused by inherited DNA differences between them. The rest of the differences are largely due to chance experiences. These environmental factors are beyond our control as parents. As we saw in the previous chapter, we don't even know what these factors are.

The second caveat is that genetic research describes the normal range of variation, genetically and environmentally. Its results do not apply outside this normal range. Severe genetic problems such as single-gene or chromosomal problems or severe environmental problems such as neglect or abuse can have devastating effects on children's cognitive and emotional development. But these devastating genetic and environmental events are, fortunately, rare and do not account for much variance in the population.

Again, parents and parenting matter tremendously, even though differences in parenting do not make a difference in children's psychological development. Parents are the most important relationship in children's lives. Still, it is important that parents get the message that children are not blobs of clay that can be moulded however they wish. Parents are not carpenters building a child by following a blueprint. They are not even much of a gardener, if that means nurturing and pruning a plant to achieve a certain result. The shocking and profound revelation for parenting from these genetic findings is that parents have little systematic effect on their children's outcomes, beyond the blueprint that their genes provide.

It is also important for parents to know that, beyond genetics, most of what happens to children involves random experiences over which parents have no control. The good news is that these don't make much of a difference in the long run. The impact of these experiences is not stable across time, as discussed in the previous chapter. Some children bounce back sooner, some later, after difficult experiences such as parental divorce, moving house and losing friends. They bounce back to their genetic trajectory.

In the tumult of daily life parents mostly respond to genetically driven differences in their children. This is the source of most correlations between parenting and children's outcomes. We read to children who like us to read to them. If they want to learn to play a musical instrument or play a particular sport, we foster their appetites and aptitudes. We can try to force our dreams on them, for example, that they become a world-class musician or a star athlete. But we are unlikely to be successful unless we go with the genetic grain. If we go against the grain, we run the risk of damaging our relationship with our children.

Genetics provides an opportunity for thinking about parenting in a different way. Instead of trying to mould children in our image, we can help them find out what they like to do and what they do well. In other words, we can help them become who they are. Remember that your children are 50 per cent similar to you. In general, genetic similarity makes the parent–child relationship go smoothly. If your child is highly active, chances are that you are too, which makes the child's hyperactivity more acceptable. Even if you both have short fuses, you can at least understand it better if you recognize your genetic propensities and work harder to defuse situations that can trigger anger. It is also useful to keep in mind that our children are 50 per cent different from us and that siblings are 50 per cent different from each other. Each child is their own person genetically. We need to recognize and respect their genetic differences.

Most importantly, parents are neither carpenters nor gardeners. Parenting is not a means to an end. It is a relationship, one of the longest lasting in our lives. Just as with our partner and friends, our relationship with our children should be based on being with them, not trying to change them.

I hope this is a liberating message, one that should relieve parents of the anxiety and guilt piled on them by parent-blaming theories of socialization and how-to parenting books. These theories and books can scare us into thinking that one wrong move can ruin a child for ever. I hope it frees parents from the illusion that a child's future success depends on how hard they push them.

Instead, parents should relax and enjoy their relationship with their children without feeling a need to mould them. Part of this enjoyment is in watching your children become who they are.

SCHOOLS MATTER, BUT THEY DON'T MAKE A DIFFERENCE

The same principles apply to education. Schools matter in that they teach basic skills such as literacy and numeracy and they dispense fundamental information about history, science, maths and culture. That is why basic education is compulsory in most countries around the world. Schools also matter because children spend half of their childhood in school.

But our focus is on individual differences. Children differ a lot in how well they do at school. How much do differences in children's school achievement depend on which school they go to? The answer is not much. This conclusion follows from direct analyses of the effect of schools on differences in students' achievement and is especially true when we control for genetic effects.

In the UK, 'league tables' rank schools by their average differences in tested achievement. In addition, rigorous government inspections of schools rank them by their quality of teaching and the support they give their pupils. Schools differ on average for both indices, but the question here is how much variance in student achievement is explained by schooling. These indices lead parents to worry about sending their children to the best schools, based on the assumption that schools make a big difference in how much children achieve.

In fact, differences in schools do not make much of a difference in children's achievement. Most striking are results using the intensive and expensive periodic ratings of school quality, including teacher quality and the atmosphere of the school, based on visits to each school every three years or so by a team of assessors from the UK Office for Standards in Education (Ofsted). We correlated these Ofsted ratings of children's secondary schools with the children's achievement assessed on the General Certification of Secondary Education (GCSE) tests administered to students in state-supported schools in England at the age of sixteen. The Ofsted ratings of school quality explained less than 2 per cent of the variance in GCSE scores after correcting for students' achievement in primary school. That is, children's GCSE scores scarcely differ as a function of their schools' Ofsted rating of quality. This does not mean that the quality of teaching and support offered by schools is unimportant. It matters a lot for the quality of life of students, but it doesn't make a difference in their educational achievement.

The conclusion that schools do not make much difference in children's achievement seems surprising, given the media attention on average differences between schools in student performance. This reflects the confusion between average differences and individual differences. Average differences between schools in the league tables mask a wide range of individual differences within schools, meaning that there is considerable overlap in the range of performance between children in the best and worst schools. In other words, some children in the worst schools outperform most children in the best schools. The biggest average difference in achievement is between selective and non-selective schools. We will look at this issue later, when we examine selection in education and occupation, which raises issues of meritocracy and social mobility.

The genetic findings reviewed in previous chapters – about heritability, non-shared environment and the nature of nurture – foreshadowed these findings. Inherited DNA differences account for more than half of the differences between children in their school achievement. Genetics is by far the major source of individual differences in school achievement, even though genetics is rarely mentioned in relation to education.

Environmental factors account for the rest of the variance in school achievement, but most of these environmental differences are not the result of systematic and stable effects of schooling. Environmental influence shared by children attending the same schools as well as growing up in the same family accounts for only 20 per cent of the variance of achievement in the school years and less than 10 per cent of academic performance at university.

The other crucial finding about the environment is the nature of nurture. What look like environmental effects are reflections of genetic differences. In relation to education, what look like environmental effects of schools on children's achievement are actually genetic effects. Examples include the correlation between student achievement and types of school and the correlation between parent and offspring educational achievement. Both correlations are usually interpreted as being caused environmentally but both are substantially mediated by genetics, as we shall see in the next chapter.

No specific policy implications necessarily follow from finding that inherited DNA differences are by far the most important source of individual differences in school achievement and that schools make so little difference. Similar to the message for parents, genetic research suggests that teachers are not carpenters or gardeners in the sense of changing children's school performance. Rather than frenetic teaching in an attempt to make pupils pass the tests that will improve their standing in league tables, schools should be supportive places for children to spend more than a decade of their lives, places where they can learn basic skills like literacy and numeracy but also learn to enjoy learning. To paraphrase John Dewey, the major American educational reformer of the twentieth century, education is not just preparation for life – education is a big chunk of life itself.

LIFE EXPERIENCES MATTER, BUT THEY DON'T MAKE A DIFFERENCE

Genetic research has far-reaching implications not just for how we think about child-rearing and schools but how we think about our own adult lives. Genetics is the major systematic influence in our lives, increasingly so as we get older. Therefore, genetics is a big part of understanding who we are. Our experiences matter a lot – our relationships with partners, children and friends, our occupations and interests. These experiences make life worth living and give it meaning. Relationships can also change our behaviour, such as helping us to stop smoking or lose weight. They can affect our lifestyle by encouraging us to exercise, play sports and go to cultural events. But they don't change who we are psychologically – our personality, our mental health and our cognitive abilities. Life experiences matter and can affect us profoundly, but they don't make a difference in terms of who we are.

This conclusion follows from the same suite of genetic findings that we have applied to parenting and schooling: significant and substantial genetic influence, the nature of nurture and the importance of non-shared environment.

Individual differences in stressful life events were among the first environmental measures for which genetic influence was found. Most research on life events used self-report measures of stressful events and their effects. However, we saw that even objectively measured events such as divorce show genetic influence. Parental divorce is the best predictor of children's divorce, but this correlation, easily interpreted as environmental, is entirely due to genetics. Quality of social support is another major aspect of life experiences that has been assumed to be a source of environmental influence but is in fact substantially caused by genetic differences.

Finding genetic influence on individual differences in 'environmental' measures led to research that showed that genetics accounts for about half of the correlations between life experiences and psychological traits, such as the correlation between perceptions of life events and depression. This is another example of the nature of nurture.

The point is that life experiences are not just events that happen haplessly to us as passive bystanders. With all our genetically rich psychological differences, we differ in our propensities to experience life events and social support. The nature of nurture suggests a new model of experience in which we actively perceive, interpret, select, modify and create experiences correlated with our genetic propensities.

The importance of non-shared environment has major implications as well for understanding why life experiences don't make a difference psychologically. The heritability of life experiences is about 25 per cent, which means that most of the individual differences in life experiences are environmental in origin. But these environmental influences are not shared by our siblings, even if our sibling is our identical twin. Our parents cannot take much credit or blame for how we turned out, other than via the genes they gave us. No one can take credit or blame because these non-shared environmental influences are unsystematic and unstable. Beyond the systematic and stable force of genetics, good and bad things just happen. As mentioned earlier in relation to parenting, the good news is that these random experiences don't matter much in the long run because their impact is not long-lasting. We eventually rebound to our genetic trajectory. To the extent that our experiences appear shared, systematic and stable, they reflect our genetic propensities. These correlations are caused genetically, not environmentally.

In summary, parents matter, schools matter and life experiences matter, but they don't make a difference in shaping who we are. DNA is the only thing that makes a substantial systematic difference, accounting for 50 per cent of the variance in psychological traits. The rest comes down to chance environmental experiences that do not have long-term effects.

Many psychologists will be aghast at this bold conclusion. Popper, mentioned earlier, said that the first commandment of science is that theories are not merely testable but falsifiable. Falsifying this conclusion is straightforward: Demonstrate that 'environmental' factors such as parenting, schooling and life experiences make a difference environmentally after controlling for genetic influence. Anecdotes are not enough, and it's not enough to show a statistically significant effect – the issue is whether these things explain more than 1 or 2 per cent of the variance. I am not worried about the conclusion being falsified, because there is a century of research behind it.

One general message that should emerge from these discoveries is tolerance for others – and for ourselves. Rather than blaming other people and ourselves for being depressed, slow to learn or overweight, we should recognize and respect the huge impact of genetics on individual differences. Genetics, not lack of willpower, makes some people more prone to problems such as depression, learning disabilities and obesity. Genetics also makes it harder for some people to mitigate their problems. Success and failure – and credit and blame – in overcoming problems should be calibrated relative to genetic strengths and weaknesses.

Going even further out on this limb, I'd argue that understanding the importance of genetics and the random nature of environmental influences could lead to greater acceptance and even enjoyment of who we are genetically. Rather than striving for an ideal self that sits on an impossibly tall pedestal, it might be worth trying to look for your genetic self and to feel comfortable in your own skin. Moreover, as we have seen, with age, as genetic influence increases, the more we become who we are.

By pointing out that most of the systematic variance in life is caused by inherited DNA differences I do not mean to imply that people should not try to work on any of their shortcomings or not try to improve certain aspects of themselves. Heritability describes *what is* but does not predict *what could be*, as I have emphasized several times. High heritability of weight does not mean there is nothing you can do about your weight. Nor does heritability mean that we must succumb to our genetic propensities to depression, learning disabilities or alcohol abuse. Genes are not destiny. You can change. But heritability means that some people are more vulnerable to these problems and also find it more difficult to overcome them.

'If at first you don't succeed, try, try, try, again' (Thomas Palmer); 'Be all that you can be' (US Army); 'Anyone can grow up to be President' (Americans) – throughout our lives we are bombarded with inspirational aphorisms like these, from childhood songs like the incy-wincy spider climbing up the water spout and stories like 'The Little Engine that Could' to adult fables like Robert the Bruce watching a spider repeatedly trying to build a web, as well as many autobiographies, novels and films about overcoming the odds. The barrage also comes from pop-psychology books whose message is that all you need to succeed is some panacea, such as the power of positive thinking or a growth mindset or grit or 10,000 hours of practice.

Anyone who is influenced by these maxims should understand that, to the contrary, genetics is the main systematic force in life. Again, this is not to say that genes are destiny. It just seems more sensible, when possible, to go with the genetic flow rather than trying to swim upstream. As W. C. Fields said, 'If at first you don't succeed, try, try again. Then quit. There's no use being a damn fool about it.'

9

Equal opportunity and meritocracy

If schools, parenting and our life experiences do not change who we are, what does this mean for society, especially for equality of opportunity and meritocracy? In particular, does it mean that the genetically rich will get richer and the poor poorer? Are genetic castes inevitable? What does this say about inequality? In this chapter, we will explore the implications of the counterintuitive findings discussed in the previous chapters.

These questions have been bound up in the topic of meritocracy, which is not the same thing as equal opportunity. Equal opportunity means that people are treated similarly, for example, everyone is given equal access to educational resources. Meritocracy only comes in when there is selection, for example, for education and employment. Meritocracy means that selection is based on capability and competence rather than unfair criteria such as wealth, prejudice or arbitrariness.

Although meritocracy sounds like an irresistibly good idea, both parts of the neologism 'meritocracy' are loaded with unpalatable connotations. The noun 'merit' refers to ability and effort but it also connotes value and worth. It is derived from the Latin word *meritum* meaning 'worthy of praise'. The '-ocracy' part of 'meritocracy' refers to power and governance. Putting these two components of meritocracy together with genetics implies that we are governed by a genetic elite whose status is justified by their ability and effort. Instead, it could be argued that people who got lucky by drawing a good genetic hand do not merit anything. Their luck at learning easily and getting satisfying jobs is its own reward. And who says we should be governed by genetic elites? The populist strain of politics around the world suggests a desire for the opposite.

The three findings from genetic research highlighted in previous chapters transform how we think about equality of opportunity and meritocracy. To reiterate, these findings are about heritability, non-shared environment and the nature of nurture. That is, genetics provides most of the systematic variation between us, environmental effects are random, and our chosen environments show genetic influence. These findings have different implications for equal opportunity and meritocracy.

At first glance, genetics seems antithetical to equality of opportunity, violating the principle enshrined in the second sentence of the 1776 United States Declaration of Independence that all people are created equal. However, the American founders did not mean that all people are created identical. They were referring to 'unalienable rights', which include 'life, liberty and the pursuit of happiness'. In less lofty terms, this means equal protection before the law and equal opportunity. But 'equal' does not mean identical. If everyone were identical, there would be no need to worry about equal rights or equal opportunity. The essence of democracy is that people are treated fairly *despite* their differences.

The most important point about equality of opportunity from a genetic perspective is that equality of opportunity does not translate to equality of outcome. If educational opportunities were the same for all children, would their outcomes be the same in terms of school achievement? The answer is clearly 'no' because even if environmental differences were eliminated genetic differences would remain.

What follows from this point is one of the most extraordinary implications of genetics. Instead of genetics being antithetical to equal opportunity, heritability of outcomes can be seen as an index of equality of opportunity. Equal opportunity means that environmental advantages and disadvantages such as privilege and prejudice have little effect on outcomes. Individual differences in outcomes that remain after systematic environmental biases are diminished are to a greater extent due to genetic differences. In this way, greater educational equality of opportunity results in greater heritability of school achievement. The higher the heritability of school achievement, the less the impact of environmental advantages and disadvantages. If nothing but environmental differences were important, heritability would be 0. Finding that heritability of school achievement is higher than for most traits, about 60 per cent, suggests that there is substantial equality of opportunity.

Environmental differences account for the remaining 40 per cent of the variance. Does this imply inequality of opportunity? To the extent that environmental influences are non-shared, this means that they are not caused by systematic inequalities of opportunity. However, as we have seen, genetic research on primary- and secondary-school achievement is an exception to the rule that environmental influences are non-shared. For school achievement, half of the environmental influence – 20 per cent of the total variance – is shared by children attending the same school. This finding implies that up to 20 per cent of the variance in school achievement could be due to inequalities in school or home environments, although this effect mostly washes out by the time children go to university.

The third finding, about the nature of nurture, is also relevant to understanding the relationship between equal opportunity and outcomes. What look like systematic environmental effects in fact reflect genetic differences. For example, the socioeconomic status of parents is correlated with their children's educational and occupational outcomes. This correlation has been interpreted as if it is caused environmentally. That is, better-educated, wealthier parents are assumed to pass on privilege, creating environmentally driven inequality in educational opportunity and stifling what is called intergenerational educational mobility.

Genetics turns the interpretation of this correlation upside down. Socioeconomic status of parents is a measure of their educational and occupational outcomes, which are both substantially heritable. This means that the correlation between parents' socioeconomic status and their children's outcomes is actually about parent–offspring resemblance in education and occupation. Phrased as 'parent–offspring resemblance', it should come as no surprise that genetics largely mediates the correlation. Parent–offspring resemblance is an index of heritability, and heritability is an index of equal opportunity. So, parent–offspring resemblance for education and occupation indicates social mobility rather than social inertia.

A more subtle way to think about the nature of nurture and its relationship to equality of opportunity is gene–environment correlation, which means that our experiences are correlated with our genetic propensities. Genetic differences in personality, psychopathology and cognitive ability make us experience life differently, as we saw in relation to the nature-of-nurture phenomenon. In relation to education, more highly educated parents provide both nature and nurture that work together to affect their children's chances to do well at school, for example, in reading and their general attitude to education. Schools select children into streams on the basis of heritable traits such as ability and previous achievement. These are examples of what behavioural geneticists call passive and reactive gene–environment correlation, respectively.

The most important type is active gene–environment correlation. Children actively select, modify and create environments correlated with their genetic propensities. For example, genetic differences in children's aptitudes and appetites affect the extent to which they take advantage of educational opportunities. This is why equal opportunities cannot be imposed on children to create equal outcomes. Genetic differences in aptitudes and appetites influence the extent to which children take advantage of opportunities. To a large extent, opportunities are taken, not given.

It would be a mistake to see gene–environment correlation as inequality, because it is, ultimately, based on genetics. For this reason, gene–environment correlation is difficult to disrupt. We can't stop parents from providing correlated nature and nurture to their children unless we adopt children away at birth. We could outlaw selection in schools, but in the classroom it is impossible as well as undesirable for teachers to treat children the same, regardless of their genetic differences. Finally, trying to stop children from actively seeking experiences correlated with their genetic appetites and abilities is futile.

What this means is that high heritability of school achievement indicates that educational opportunities are substantially equal. Attempts to increase equality of opportunity should focus on reducing shared environmental influence, although shared environment at most accounts for 20 per cent of the variance in school achievement. Non-shared environmental influences are out of reach because they are unsystematic and we don't know what they are. Correlations between opportunity and outcome are genetically driven. This is another way in which DNA makes us who we are.

It is worth reiterating that this genetic research describes the mix of genetic and environmental influences on individual differences in school achievement in specific samples at specific times. Most of the research comes from developed countries, especially Europe and the US, in the twentieth century. The results could be different for different countries in different times. Our focus here is on the effects of equal opportunity on individual differences in school achievement. As access to education broadens, heritability would be expected to increase. The first twin study on this topic found that heritability of educational attainment increased and the impact of shared environment decreased in Norway following the Second World War, when access to education expanded. Subsequent studies in several countries also found increased heritability and decreased shared environmental influence after the Second World War, as equality of educational opportunity increased. Some recent evidence suggests this might be going in reverse in the US in the twenty-first century, with decreased heritability and increased shared environmental influence on educational attainment, which suggests there is greater inequality of educational opportunity.

In contrast to equal opportunity, the concept of meritocracy is relevant only when there is selection, for example, selecting children into certain schools. At the level of primary school in the UK there is little selection because most parents send their children to a local school. Equal opportunity in this case means that children at different schools receive equally good education.

Selection becomes more of an issue at the level of secondary school. Students vie to get into the 'best' secondary schools, which leads to selection. The issue of meritocracy is about the extent to which selection is based on 'merit' – in this case, on the students' ability, prior achievement and other predictors of success.

In the UK the biggest average difference in student achievement is between state-funded non-selective schools, or comprehensive schools, and selective schools, which include state-funded grammar schools and privately funded schools. The average GCSE scores for children in selective schools, whether grammar or private, are a whole grade higher than for children in non-selective schools.

This average difference in achievement between selective and non-selective schools has been assumed to be caused environmentally – selective schools are assumed to provide better schooling. However, genetic research shows that this difference cannot be credited to better education in selective schools. By definition, selective schools select the most competitive students, choosing meritocratically on the basis of students' prior achievement and ability and, less meritocratically, on family wealth. For example, at the top secondary schools students are interviewed and tested for several years before they are admitted. In addition, parents and students select the 'best' secondary schools in part on the basis of these same factors. That is, if students have not performed well on tests of school achievement in primary school, they are not likely to aspire to high-flying secondary schools.

So it should come as no surprise that students in selective schools perform better than students in non-selective schools, because it is a self-fulfilling prophecy that the students selected by selective schools for their school achievement have higher GCSE scores. When we control for the factors that are used to select students the average difference in GCSE scores is negligible and overall GCSE variance explained by school type shrinks to less than 1 per cent. In other words, selective schools do not improve students' achievement once we take into account the fact that these schools preselect students with the best chance of success.

This is another example of gene–environment correlation, in that students select schools and are selected by a school in part on the basis of the students' prior school achievement and ability, which are highly heritable. This explains what would otherwise appear to be an odd result, which we will examine later: Students in selective and non-selective schools differ in their DNA. Because the traits used to select students are highly heritable, selection of students for these traits means that students are unintentionally selected genetically.

If better achievement by students in selective schools than by those in non-selective schools were due to value added by selective schools, this would imply inequality of educational opportunity. But because the difference in achievement disappears after controlling for selection factors, we can conclude that selection is meritocratic. For this same reason, differences in GCSE results for selective and non-selective schools are not an index of the quality of education the schools provide. An attempt to create a fairer comparison was implemented in England in 2017 by correcting GCSE scores at the end of secondary school for achievement at the end of primary school at the age of eleven. This innovation was sold as an index of the value added by schools, which is called 'progress'. However, we have found that this measure of 'progress' is still substantially heritable (40 per cent), which means that it is not a pure index of students' 'progress' or schools' added value. How is it possible that this measure of 'progress' is so heritable? The answer is that correcting for school achievement at the age of eleven does not correct for other heritable contributions to performance on the GCSE test such as intelligence, personality and mental health.

Even though schools have little effect on individual differences in school achievement, some parents will still decide to pay huge amounts of money to send their children to private schools in order to give their children whatever slight advantage such schools provide. Even for state-supported selective grammar schools, some parents who can afford to do so will pay a premium to move house to be within the catchment of a better school. I hope it will help parents who cannot afford to pay for private schooling or move house to know that it doesn't make much of a difference in children's school achievement. Expensive schooling cannot survive a cost–benefit analysis on the basis of school achievement itself.

There may be benefits of grammar and private schools in terms of other outcomes, such as better prospects for university, making connections that lead to job opportunities later in life, and imbuing students with greater confidence and leadership skills. For example, although only 7 per cent of students in the UK attend private schools, their alumni notoriously dominate the top professions – over a third of MPs, over half of senior medical consultants, over two-thirds of high court judges and many top journalists.

But are these advantages merely another example of the

self-fulfilling prophecy of selecting the best students in the first place? In the case of the difference in GCSE scores between selective and non-selective secondary schools, we have seen that the difference disappears after controlling for factors used in selection. We have found similar results for university prospects. That is, students from selective secondary schools are much more likely to be accepted by the best universities, but this benefit largely disappears after controlling for selection factors. In other words, the students would have been as likely to be accepted by the best universities if they had not gone to a selective secondary school. Indeed, changes in selection criteria for the best universities actually favour a student who does well at a comprehensive secondary school.

It seems likely that the other potential advantages of selective schooling – such as occupational status, income and personal characteristics – are also self-fulfilling prophecies rather than value added by selective schools. Finally, it should be emphasized that if all secondary schools were equally good, there would be no need to select students in the first place. If there were no selection, there would also be a lot less stress for students and their parents. In addition, neighbourhood schools foster social integration and a sense of community.

We have used education as an example of the links between opportunities, capabilities and outcomes, but the same issues apply to occupational status and income. Here, as long as getting a high-status job and making lots of money are priorities, selection is necessary, which raises the issue of the criteria used for selection. As in the example of the over-representation of private schooling among MPs, medical consultants and high court judges, is selection for occupational status and income based on advantage or ability?

Both occupational status and income are substantially heritable, about 40 per cent in more than a dozen twin studies in developed countries. This should not be surprising, because occupational status and income are related to educational attainment and intelligence, which are heritable traits. Similar to the argument we made for education, heritability is an index of meritocratic selection for occupational status and income, so we can conclude on the basis of substantial heritability that selection is considerably meritocratic. Unlike education, shared environmental influence for occupational status is negligible, which means that environmental influences are random and that most of the systematic effects on occupational status and income can be attributed to genetics.

Anyone who has interviewed candidates for a job knows the complexity and capriciousness of selection. In the first place, you can only select from people who applied for the position. In addition, interviews are notoriously poor predictors of performance. These and many other unsystematic factors, including chance, contribute to individual differences in occupational status and income. These factors are not meritocratic, but they do not represent systematic bias.

The nature-of-nurture issue is also relevant for occupations. What look like systematic environmental effects are reflections of genetic effects. An important example is the similarity between parents and their offspring in occupational status and income. As examined earlier in relation to education, parent–offspring resemblance for occupational status and income cannot be assumed to arise from environmental advantages passed on from parent to child. The correlation is chiefly caused genetically, which indicates that the systematic effects of selection, including self-selection, are substantially meritocratic. The same is probably true for the ostensible effect of private schooling on occupational success, as noted earlier.

I would argue that anything that increases the heritability of occupational status and income makes the selection process more meritocratic. The absence of shared environmental influence implies that there are few systematic environmental inequities in the population as a whole, which means that environmental levers for change are not within our grasp. Inherited wealth, which is the epitome of inequity, can be changed, for example, by taxing wealth rather than income. However, inherited wealth is not much related to occupational status or to income, at least as income is currently defined by tax authorities. So, tackling inherited wealth will not make much difference in occupational status or income per se. One thing that would make a difference is to make selection processes more effective in predicting performance, because this would reduce unsystematic influences on occupational status and income. The DNA revolution will transform the selection process by introducing the most systematic and objective predictor of performance by far: inherited DNA differences.

At first thought, it might seem that, given free rein, genetics will limit social mobility and calcify society into genetic castes, as happened in India, where for thousands of years mating was limited to members of the same caste. I would argue that this is not a problem in modern societies for two reasons. The first is simple: a lot of the environmental variation between us is not systematic. Random effects will not create stable castes.

The second reason is that parents and offspring are only 50 per cent similar genetically. Their genetic similarity means that, on average, brighter parents have brighter children. But their 50 per cent genetic dissimilarity means that children of brighter parents will show a wide range of ability, including some children of lower-than-average ability. If you take pairs of individuals randomly, their average difference will be seventeen IQ points. First-degree relatives – parents and their offspring or siblings – differ by thirteen IQ points on average. This allows plenty of room to go down as well as up the ladder.

In addition, children of high-IQ parents will on average have lower IQ scores than their parents for the same reason that tall parents have taller-than-average children but those children are less tall than they are. For the same reason, most prodigies do not have prodigy parents. This is a statistical phenomenon, not a specific genetic process. That is, the same phenomenon would occur if individual differences were due to systematic environmental factors indexed as shared environment. However, genetics, not shared environment, is the systematic source of individual differences, and it is genetics that leads to concerns about castes.

If children were genetically unrelated to their high-IQ parents, as is the case for adopted children and their adoptive parents, the children's mean IQ would be expected to be 100, if the adopted children were representative of the population. Because children are 50 per cent similar genetically to their parents, genetics predicts that the children's average IQ will regress halfway from their parents' IQ to the population average. For example, parents with an average IQ of

130 are expected to have children whose average IQ is 115, regressing halfway back to the population average of 100. This reshuffling of DNA differences in the genetic lottery prevents the evolution of a rigid genetic caste system.

The flip side of this argument is that parents of average ability also have children with a wide range of ability, including children of high ability. Because there are many more parents of average ability than of high ability, this guarantees that most of the individuals of highest ability in the next generation will come from parents of average ability, not from the most able parents. As long as downward social mobility as well as upward social mobility occurs, we do not need to fear that genetics will lead to a rigid caste system.

Even though most of the systematic differences between people are genetic in origin, this does not mean that we need to be fatalistic and accept the status quo. One reason, emphasized earlier, is that genetics describes what is – it does not predict what could be. You can beat the genetic odds. But it is not fatalistic to recognize that DNA matters and to appreciate genetic differences between our children and between ourselves. It seems only sane to suggest that, when you can, try to go with the grain of genetics rather than fight against it.

A second way to avoid fatalism is to deny the value system that drives the debate about meritocracy and social mobility. It assumes that the point of education is to get better test scores in order to get a better occupation and that the point of an occupation is to achieve high status and make lots of money. Another way of looking at education is as a time to learn basic skills and to learn how to learn and to enjoy learning. It is a decade of their lives when children can find out what they like to do and what they are good at doing, where they can find their genetic selves, which may not dispose them towards higher education. Everyone should be given the chance to learn at school, but not everyone will choose (or can afford) to go on to university.

Similarly, with occupations, where selection cannot be avoided, we will end up with a lot of frustrated people if we only value high-status occupations. Society needs people who are good care workers, nurses, plumbers, janitors, policemen, mechanics and public servants. What

I want most for my children is that they are happy and that they are good people. It would be a terrific bonus if they like what they do.

Self-selection is an important factor to the extent that people are free to choose what they do to earn a living. Self-selection involves listening to genetic whispers, not just about intelligence but also about personality and interests. These options include choosing a job that just pays the bills rather than a high-income occupation that might come with a high-stress price tag, or an especially enjoyable vocation that might not pay the bills. Beyond the money needed to get by, letting money define success in life does not achieve happiness, enjoyment or goodness. In a just society, jobs that require less 'merit' would nonetheless be rewarded monetarily so that they provide a reasonable standard of living.

We could also deny the value system based on money at a more political level. Much of the concern about inequality and social mobility is about income inequality. Individual differences in income are, like everything else, substantially heritable, about 40 per cent. Income correlates with intelligence, and genetics drives this correlation. But this does not mean that higher intelligence merits more income. I would argue that genetic wealth is its own reward. If society really wanted to reduce income inequality, it could do so directly and immediately with a tax system that redistributes wealth.

My value system suggests that we need to replace meritocracy with a just society. Although rigid genetic castes will not come into being, social mobility creates genetic inequality, which leads to an inherent inequality of opportunity. That is, children dealt a lucky genetic hand have a better chance of doing well at school and getting a better job and making more money. This inequality in outcome is not going to be tackled indirectly through the educational system. As mentioned, if all children were taught exactly the same, their genetic differences would still lead to differences in their achievement, which would lead to differences in occupational outcomes. Again, economic inequality could be dealt with directly through a redistributive tax system that reduces the gap between rich and poor.

I think people are more concerned with fairness and a just society than with economic inequality per se. It seems unfair that 60 per cent of the increase in US national income in the last three decades went to just the top 1 per cent of earners, primarily due to soaring salaries at the top end of the pay scale. However, I would argue that more important than the relative inequality of income for this top 1 per cent is the absolute inequality of the bottom third, whose debts exceed their assets.

Equality of opportunity, income inequality and social mobility are some of the most critical issues in society today. They are hugely complicated topics that heavily depend on values. My objective was to look at these issues through the single lens of genetics, to show how DNA makes us who we are. However, no specific policies necessarily follow from genetic findings, because policies depend on values. My values, not my science, lead me away from meritocracy towards a just society.

The DNA revolution will make all these genetic implications more personal because we will be able to predict genetic risk and resilience, strengths and weaknesses, for individuals. The second part of *Blueprint* explores the DNA revolution and its implications for individuals, psychology and society.

PART TWO

The DNA revolution

10

DNA: The basics

In order to grasp the significance of the DNA revolution and how DNA makes us who we are, it is important to understand a few basic facts about the blueprint for life. I'm sorry, therefore, if this chapter occasionally seems like a biology lesson, but it only describes the essentials needed for DNA literacy, especially in relation to understanding the DNA revolution as it affects psychology. The single most important thing to know is that DNA consists of dumb molecules that blindly obey the laws of chemistry. Together, these molecules, which are the same in each of our trillions of cells, produce life in all its amazing complexity.

In 1866 Gregor Mendel showed how heredity works functionally. Mendel carefully fertilized thousands of pea plants over many years in the garden of his monastery in what is now the Czech Republic. On the basis of his many experiments with traits such as whether the seed had smooth or wrinkled skin, Mendel concluded that there are two 'elements' of heredity for each trait in each individual and that offspring receive one of these two elements from each parent.

Until the 1950s it was still a mystery as to what these 'elements' were. In 1953 James Watson and Francis Crick described the famous double-helix structure of DNA, which beautifully filled the bill for Mendel's elements. The double helix consists of two strands coiled around each other (Figure 3).

DNA is like a rope ladder with the two strands of rope held together by weak, easily broken rungs. The double-helix shape comes from twisting the rope ladder so that it forms a spiral. The two strands of the rope ladder are weakly held together by rungs that consist of chemical bonds between four molecules called nucleotides: A

Figure 3 The double helix of DNA

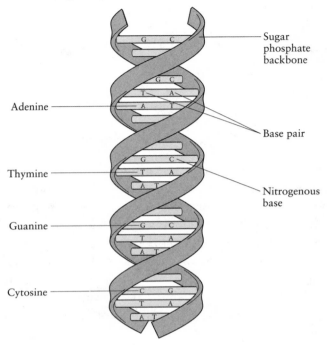

Sugar
phosphate
backbone

Adenine

Base pair

Thymine

Nitrogenous
base

Guanine

Cytosine

(adenine), C (cytosine), G (guanine) and T (thymine). The backbone of the double helix consists of alternating sugar and phosphate molecules. This sugar phosphate backbone and the nucleotide rungs gave DNA its name: deoxyribonucleic acid.

In a paper that was just over two pages long yet still the most important ever produced in biology, Watson and Crick wrote that 'the sequence of bases on a single chain does not appear to be restricted in any way'. In other words, looking at one strand of the rope ladder, you can see any sequence of A, C, G and T, which suggested that the genetic code could lie in each strand's sequence of A, C, G and T nucleotides.

In 1961 Francis Crick and Sydney Brenner began to crack the genetic code by showing that the code consists of a sequence of three rungs on the rope ladder (e.g., A-A-A or C-A-G or G-T-T), which is like a three-letter 'word'. The four letters (A, C, G, T) taken three at a time yield sixty-four possible combinations. In the next few years the meaning of all sixty-four words in the DNA dictionary was gradually worked out. For example, A-A-A is one word, C-A-G is another and G-T-T is another. These words code for one of twenty amino acids. There are hundreds of amino acids but only twenty are produced from scratch by our DNA. For example, A-A-A codes for phenylalanine, C-A-G for valine, G-T-T for glutamine. Some three-letter words code for the same amino acid and some provide punctuation such as start and stop signals, using up all sixty-four words in the DNA dictionary.

Why amino acids? Amino acids are the building blocks of proteins, which are integral to all that we are. Proteins are essential for the structure, function and regulation of our bodies, including neurons and neurotransmitters, which are the basic elements of our brain and who we are psychologically. The average protein contains a unique sequence of the 20 amino acids, varying from 50 to 2,000 amino acids in length. With 20 amino acids in any order in such long strings, there is a limitless variety of proteins. On average, each of our cells produces 2,000 different proteins.

The strands of the double helix are held together by weak chemical bonds between the A, C, G and T nucleotides. These four molecules produce only four types of rungs, not all twelve possible rungs. The reason is that A bonds only with T and G bonds only with C. So, there are only four types of rungs in the rope ladder of DNA: A-T, T-A, C-G and G-C, as shown in the schematic figure of DNA above.

This model of DNA coding for amino acids is what the word 'gene' classically meant. However, we now know that DNA does much more than code for amino-acid sequences. Only 2 per cent of the human DNA sequence works like this; there are only 20,000 classical 'genes'. The other 98 per cent of DNA was thought to be junk but is now known to have important functions, as I will describe later.

In a classic example of understatement, Watson and Crick wrote, 'It has not escaped our notice that the specific pairing we have postulated immediately suggests a possible copying mechanism for the genetic material.' What they meant was that, if the two strands of the double helix are unzipped, each strand's sequence of nucleotide bases of A, T, C and G would seek its complementary mate (A with T; T with A; C with G; G with C). This would result in the creation of two identical DNA double helices. Those two cells create four cells, eight cells, then sixteen cells, and so on. It neatly provides a mechanism for explaining how we begin life as a single cell and end up with 50 trillions of cells, each with the same DNA.

We have 3 billion rungs in the double helix of DNA, which is called the genome. But the genome is not one continuous rope ladder with 3 billion rungs. It is broken up into twenty-three segments, or chromosomes, which vary in length from 50 million to 250 million rungs.

We actually have 6 billion nucleotide bases because our DNA blueprint consists of two genomes, one from our mother and the other from our father, as Mendel deduced from his pea-plant experiments. So, we have twenty-three *pairs* of chromosomes – one member of each pair comes from the mother's egg and the other from the father's sperm. Egg and sperm cells are the only cells to have just one chromosome from each pair, so when an egg and sperm combine they produce a cell that has a full set of paired chromosomes. This single cell divides to create two cells, and they each divide again and again, resulting in the trillions of cells in our bodies, each with the same DNA sequence.

It is random which of your mother's pair of chromosomes you received for each of the twenty-three chromosomes, and similarly for your father. For each pair of chromosomes, your sibling has a fifty-fifty chance of getting the same chromosome as you, which is why siblings are, on average, 50 per cent similar. The exception is identical twins, who have exactly the same chromosomes because they come from the same fertilized egg. This is why siblings are similar but also different in terms of psychological traits and why identical twins are more similar than other siblings.

About 99 per cent of the 3 billion rungs in the DNA sequence are identical for you and me. This DNA is what makes us similar. But that means there are 30 million rungs that differ between us. As we have seen, these differences in DNA sequence are the blueprint that makes us what we are.

As new cells are formed, the double helix unzips and each strand of the rope ladder seeks its complement for each rung. This duplication process is incredibly reliable, but mistakes are made – *mutations* – which are like typos in the genetic code. When a mutation occurs in egg or sperm, it is passed on to offspring, who then pass it on to their offspring.

All kinds of differences in DNA sequence can occur, but the most common is when a single rung differs between people. A change in one of the 3 billion rungs in the double helix of DNA is called a single-nucleotide polymorphism (SNP, pronounced 'snip'). You and I have about 4 million SNPs but many of these are present only in a few people, which means that we do not have the same 4 million SNPs. There may be as many as 80 million SNPs in the world. Any particular population – the UK, for example – has about 10 million SNPs. The rest of this book focuses on SNPs, because they have played a central role in the DNA revolution.

All we inherit is the DNA sequence in the single cell with which our lives began, with its unique combination of DNA differences. Although all cells have the same DNA, cells express only a small portion of all DNA. Different types of cell – for example, brain, blood, skin, liver and bone cells – express different bits of DNA. DNA sequence is transcribed by a messenger molecule called RNA. RNA is then translated into amino-acid sequences according to the genetic code. This process is what is meant by the term *gene expression*.

Many mechanisms affect gene expression. Some are long-term mechanisms (called *epigenetic*) that involve adding molecules to the DNA that prevent its transcription. Other mechanisms for expression have shorter-term effects. For example, proteins that interact with DNA regulate transcription in response to cues from the environment. You are changing the expression of many genes that code for neurotransmitters in your brain as you read this sentence. As the neural processes involved in reading deplete these neurotransmitters, you express the genes that code for these neurotransmitters in order to replenish them.

If two individuals differ in their DNA sequence, a SNP, for example, that codes for a particular neurotransmitter, that SNP will be faithfully transcribed when that stretch of DNA is expressed. This DNA difference could be translated into different amino-acid sequences for the two individuals. This change in amino-acid sequence could alter how well the neurotransmitter works. The key point is that all we inherit is DNA sequence. Gene expression does not change our inherited DNA sequence. If a SNP is associated with a psychological trait, that means the SNP was expressed.

Let's zoom in on one of the 10 million SNPs in the human genome.

For reasons that will become clear, let's focus on one SNP that happens to be in the middle of chromosome 16. Chromosome 16 has 90 million rungs in the double helix, and this SNP is at rung number 53,767,042. This mutation could have been A, C, T or G – but it happened to be T, until a mutation occurred long ago that changed T to A in one individual. The person with this mutation passed on this new A nucleotide to half of their offspring, who then passed it on to half of their offspring. After several generations, the new A nucleotide spread in the population. Perhaps its frequency increased because it conveyed some slight advantage evolutionarily, which is the case for this particular mutation, as we shall see. More often, its frequency increased because it didn't have any effect and it just spread from generation to generation, following Mendel's laws of inheritance. Today 40 per cent of all chromosomes have the A nucleotide at this spot on chromosome 16. The other 60 per cent has the original T nucleotide. These alternate forms of DNA sequence are called *alleles*.

Because we inherit a pair of chromosomes, one from each parent, we have one allele from each parent. The pair of alleles is called our genotype. For the SNP on chromosome 16, we could inherit either an A allele or a T allele from our mother and an A or a T from our father. If we inherit an A allele from both parents, our genotype is AA. If we inherit an A from one parent and a T from the other, our genotype is AT. The third possibility results in a TT genotype. For this spot on chromosome 16, 15 per cent of us are AA, 50 per cent are AT and 35 per cent TT. Genotypes are just alleles considered two at a time, the way they are packaged in individuals. If you count the alleles in these genotype frequencies, you get the allele frequencies of 40 per cent A and 60 per cent T.

The reason for focusing on this particular SNP is that it was one of the first SNPs found to be associated with a complex trait, in this case body weight. Each A allele is associated with a three-pound increase in body weight. Adults with AT genotypes weigh three pounds more on average than people with TT genotypes, and people with AA genotypes weigh three pounds more than those with AT genotypes. We can correlate these genotypes with weight by giving everyone a score based on the number of A alleles they have: 0 for TT genotypes, 1 for AT genotypes, and 2 for AA genotypes. This correlation in European populations is 0.09, which accounts for less than 1 per cent of the differences in weight between people. The heritability of weight is 70 per cent, so this SNP association explains only a tiny portion of the heritability of weight.

How does this SNP work? The SNP is in a gene called *FaT* mass and Obesity-associated protein, which is mercifully shortened to the acronym *FTO* (rather than succumbing to the temptation to call it *FATSO*). The *FTO* gene codes for a type of protein called an enzyme which accelerates chemical reactions. The *FTO* enzyme affects gene expression, the basic process by which DNA is transcribed into RNA. The *FTO* gene comprises half a million A, C, T and G rungs in the middle of the 90 million rungs on chromosome 16. Our target SNP is about 100,000 rungs up the 500,000 rungs of this *FTO* stretch of chromosome 16.

Mutations can change the meaning of the three-letter words of DNA. For example, as mentioned earlier, the three-letter sequence C-A-G codes for the amino acid valine. If the C were changed to G, the three-letter code would be G-A-G, which would code for the amino acid leucine instead of valine. Changing just one amino acid in the chain of hundreds of amino acids that create a protein can drastically alter the function of the protein. Thousands of disorders are caused by mutations in the genetic code that change the amino-acid sequence of proteins. Many such mutations are lethal.

The possibility of actually correcting a DNA mutation has been realized recently. A gene-editing technique called CRISPR (Clustered Regularly Interspaced Short Palindromic Repeats, pronounced 'crisper') can efficiently and precisely cut and replace a DNA mutation. CRISPR has led to many advances in understanding how genes work. Its most exciting but controversial feature is that it can be used to correct a DNA mutation in embryos, whose offspring would also be free of the mutation. Ethical concerns about unintended consequences of permanently changing the human genome in this way limit the use of CRISPR in embryos. Researchers are attempting to use CRISPR to treat several single-gene diseases in somatic cells that are not passed on, including muscular dystrophy, cystic fibrosis and certain blood diseases. The problem is that, unlike changing the DNA in an embryo, which has just a few cells, or in sperm or an egg, which are single cells, DNA needs to be edited in many cells in blood or muscle or lung to bring about a therapeutic effect. In contrast, genetic influence on psychological traits is not a matter of a hard-wired single-gene mutation. Heritability is the result of thousands of genes of small effect. For this reason, gene-editing seems unlikely to be used to alter genes involved in psychological traits.

In fact, our SNP is in a stretch of DNA in the *FTO* gene that does not code for proteins. It turns out that less than 2 per cent of the genome's DNA sequence codes for proteins. These are the 20,000 classical genes mentioned earlier. Most mutations are in the other 98 per cent of DNA that does not code for a change in amino-acid sequence and used to be called 'junk DNA' because it is not translated into amino-acid sequences. Even within genes like the *FTO* gene, most of the DNA does not code for proteins. These non-coding stretches of genes, or *introns*, are spliced out of the RNA code before the RNA is translated into proteins. The remaining RNA segments, or *exons*, are spliced back together and proceed to be translated into amino-acid sequences.

We are still learning about the many ways in which mutations in these non-coding differences in DNA sequence make a difference. What we do know is that they do make a difference. Some research suggests that as much as 80 per cent of this non-coding DNA is functional, in that it regulates the transcription of other genes. This distinction is important because most DNA associations with psychological traits involve SNPs in non-coding regions of DNA rather than in classical genes.

The general answer to the question of how this *FTO* SNP affects weight is the same as the answer for thousands of such SNP associations with traits throughout biological and medical science: it's complicated. This is not glib – it is an important discovery about how DNA differences affect complex psychological traits. Natural selection did not tinker with the genome to make things simple for scientists. The *FTO* SNP that is associated with body weight does not operate in a straightforward way to affect, for example, some single metabolic process. Pathways between genes and complex traits are difficult to trace because each SNP has many different effects (pleiotropy) and each trait is influenced by many SNPs (polygenicity), as mentioned earlier. These two principles are key to understanding the DNA revolution in psychology. Pleiotropy and polygenicity mean that many DNA differences of small effect are likely to affect psychological traits – which is the case, as we shall see.

The question of how DNA affects behaviour can be addressed at many levels, for example at the level of biochemistry, physiology, neurology and psychology. Biologists like to find answers to the question 'how' at the biochemical level so that knowledge about the *FTO* SNP, for example, can be translated into a weight-loss pill. The *FTO* SNP alters the expression of several genes in fat cells, affecting how much fat they store away in reserve. For people with the AA genotype, these genes are more easily turned on, telling the fat cells to stock up on fat. If we could figure out how the AA genotype does this, it might suggest how to stop this process and reduce weight, although there is always concern about the unintended consequences of altering highly polygenic and pleiotropic systems, with their evolved checks and balances.

The A allele probably spread throughout the population because the mutation was advantageous early in the evolution of our species. Individuals with the A allele stored up extra fat. This extra fat could have saved them from starving when the next meal was days away. The problem for us today is that we have a Stone Age brain in a fast-food world with easy access to high-energy foods. Today, we don't need the A allele to help us store extra fat – that A allele is now a liability.

In contrast to the bottom-up approach of biologists, psychologists take a top-down approach to the question 'how?' by trying to find answers at the level of behaviour rather than of biochemistry. In the case of the *FTO* SNP associated with weight, we discovered that the A allele increases responsiveness to food cues and decreases the extent to which we feel full after eating, or satiety. Psychologists are happy to find behavioural explanations because these explanations can suggest low-tech, cost-effective behavioural interventions. For example, finding that this SNP affects satiety suggests that satiety-related behavioural interventions could be effective in losing weight. That is, we can learn to pay more attention to feeling full in order to counteract the A allele's effect, especially for people with AA genotypes.

Finding associations between SNPs and complex traits like the association between the *FTO* SNP and weight in the past decade marked the beginning of the DNA revolution.

*

How are SNPs genotyped? There are three steps in the process: getting cells, extracting DNA from the cells and genotyping the DNA. (If you would like to know more about these steps, see the Notes section at the end of the book.) So far, we have examined genotyping a single SNP, using the *FTO* SNP as an example. We all have millions of SNPs in our genome. Genotyping them one by one would cost many millions of pounds just to genotype one individual for all the SNPs in the genome.

Because each of us has two genomes, one from each parent, we have 6 billion nucleotide bases in our genome. If we knew the sequence of these 6 billion bases for many individuals, we could identify all of the inherited DNA differences, not just SNPs, that make a difference in psychological traits. This is now happening; it is called *whole-genome sequencing*. Rather than 'just' genotyping millions of SNPs, whole-genome sequencing works out the sequence of all 6 billion nucleotide bases. As noted earlier, 99 per cent of the 6 billion rungs in the DNA sequence are identical for you and me. But that means there are 30 million rungs that can differ between us. Remember that we are interested in these DNA differences because it is the differences that make us different. Whole-genome sequencing can identify all of these DNA differences. The genome sequence is the end of the story – that's all we inherit.

The first human genome sequence was completed in 2004, based on the work of hundreds of scientists for a decade and costing more than £2 billion. Today, a human genome of 6 billion nucleotide bases can be sequenced in a day for less than £1,000.

However, the DNA revolution began about a decade ago with a different technological advance that became possible thanks to us knowing the sequence of the whole genome. Looking at the whole genome sequence for many individuals began to reveal millions of DNA differences, including SNPs. Rather than laboriously and expensively sequencing the whole genome of individuals, SNP *microarrays* were developed that focused on genotyping SNPs rather than sequencing the entire genome.

SNP microarrays are often called SNP chips because they are analogous to the silicon chips at the heart of your computer. SNP chips use the traditional process to genotype SNPs. But instead of genotyping SNPs one at a time, the chip, which is the size of a postage stamp, simultaneously genotypes hundreds of thousands of probes for DNA sequences throughout the genome.

As a first step in screening the genome for SNP associations, it is not necessary to genotype each of the millions of SNPs throughout the genotype. Many SNPs are very close together on a chromosome and are inherited together as a package. In other words, if you know an individual's genotype for one SNP, you know their genotype for the other SNP. For this reason, a SNP chip that genotypes a few hundred thousand SNPs strategically chosen can capture information about most of the common SNPs in the genome. Common SNPs are those with allele frequencies greater than 1 per cent in the population. For example, our *FTO* SNP has an allele frequency of 40 per cent for the A allele and 60 per cent for the T allele. The fact that SNP chips genotype only common alleles turns out to be important later in this story.

SNP chips are now cheap, costing less than £50, and have been used to genotype millions of people for hundreds of thousands of SNPs across the genome. Until SNP chips became available, attempts to find DNA differences associated with psychological traits were limited to laboriously genotyping SNPs in a few 'candidate' genes thought to be important for a particular trait. This candidate-gene approach did not pay off and led to many false positive findings that did not replicate, as we shall see in the next chapter.

SNP chips made it possible to scan the entire genome to identify SNPs associated with complex traits and common disorders, rather than just looking at a few candidate genes. This systematic approach is called *genome-wide association* (GWA). Genome-wide association studies kicked off the DNA revolution by providing the first effective tool to hunt for genes responsible for the heritability of psychological traits. We will join the hunt in the next chapter.

The goal of this chapter is to provide the essential elements of DNA literacy, especially in relation to the DNA revolution in psychology. This includes the structure and function of the double helix of DNA, the genetic code, mutations in the genetic code, a particular type of mutation called a SNP, gene expression, genotyping SNPs, and SNP chips. These are the ingredients of the DNA revolution.

11

Gene-hunting

The first law of behavioural genetics is that all psychological traits show a significant and substantial genetic influence. Heritability means that inherited differences in DNA sequence cause differences between us. This chapter is about the hunt for these DNA differences, made possible for the first time by SNP chips. Nothing would advance the genetics of psychological traits more than finding these DNA differences because it would make it possible to predict psychological traits for individuals directly from DNA. Prediction from DNA differences will have a huge impact on psychology, on society and on you, as will be seen in the rest of this book.

The hunt for the genes responsible for the pervasive heritability of psychological traits began in earnest about twenty-five years ago. After several false starts and surprises there have been dramatic breakthroughs in the last two years. To appreciate these breakthroughs, which signal the dawn of the DNA revolution, we pick up the story as I experienced the chase in relation to my research on cognitive abilities and disabilities. For two decades the hunt to find the DNA differences responsible for the heritability of these traits was getting nowhere, despite promising new techniques. I almost gave up at one point. Finally, the quest closed in on its quarry, but the shock was to find that the quarry was not the big game we had set out to find.

When the hunt began twenty-five years ago everyone assumed we were after big game – a few genes of large effect that were mostly responsible for heritability. For example, for heritabilities of about 50 per cent, ten genes each accounting for 5 per cent of the variance would do the job. If the effects were this large, it would require a sample size of only 200 to have sufficient power to detect them.

This was wishful thinking because, at that time, before SNP chips, each individual had to be genotyped one SNP at a time. Genotyping one SNP at a time was extremely slow and expensive. As a result, only a few SNPs in a few candidate genes were genotyped for a few hundred individuals. For psychological traits, the obvious candidate genes were those that affected brain neurotransmitters. Hundreds of brain-related genes have been the focus of candidate-gene studies of psychological traits during the last two decades. The euphoria of beginning to find genes that predict psychological traits came crashing down as it became clear that none of these reported associations replicated. This fiasco was genetics' contribution to the replication crisis described earlier. (If you are interested in reading more about the candidate-gene fiasco, please see the Notes section.)

The pain of this false start was eased by the success of a new approach that came about at the turn of the century, just as it was becoming clear that candidate-gene studies were a flop. The new approach was genome-wide association, which is the opposite of the candidate-gene approach. The dream was to look systematically across the genome rather than picking a few, somewhat arbitrary, candidate genes. To do this would require tens of thousands of SNPs genotyped for each of thousands of individuals. Although genotyping costs had gone down by then, it still cost about ten pence to genotype one DNA marker for one individual. So, genotyping 'just' 10,000 DNA markers one by one for 1,000 individuals would cost almost £1 million and a lot of time.

I didn't have a £1 million for such a study but in 1998 I decided to screen the genome, genotyping DNA differences one by one, in order to find DNA differences associated with intelligence, using a couple of tricks to reduce the expense and time to do it. Despite these shortcuts, the study took two years to complete. The results, published in 2001, were very disappointing, a second false start. Although we had power to detect associations that account for 2 per cent of the variance in intelligence, not a single association with intelligence survived our stringent replication design. Taken at face value, these results implied that DNA associations with intelligence account for less than 2 per cent of the variance.

But it was more comforting not to take these results at face value.

I had many technical reasons not to trust the results in this uncharted territory, but the main reason not to believe them was the direct implications if the results were true. Huge amounts of time and money would be needed to detect such small effect sizes, and even if we committed the resources needed to meet these daunting challenges, there was no guarantee that it would pay off.

In the early 2000s the SNP chip began to appear, which made genome-wide association studies tremendously easier and less expensive because chips could genotype many SNPs for an individual quickly and inexpensively. SNP chips triggered the explosion of genome-wide association studies.

I was excited about this technological advance and pounced on the first SNP chip. This chip had only 10,000 SNPs and cost £400 per person, which is ten times more expensive than current SNP chips that genotype hundreds of thousands of SNPs. I used these chips to try to find SNP associations with intelligence in my TEDS sample of 6,000 UK children. But again, the results were very disappointing. The biggest effects accounted for only 0.2 per cent of the variance of intelligence and did not replicate. I was beginning to think my luck had run out – after a decade of work, this was the third false start.

What these results were trying to tell us, just like my previous study, was that the biggest effects are much smaller than we thought. It felt like the cartoon about a scientist with a smoking test tube who asks a colleague, 'What's the opposite of Eureka?' It was very hard to believe that genetic effects are so small. Again, it was easier to think that something must be wrong with my studies. Believing in these results meant that the heritability of intelligence – and probably all psychological traits – is caused by thousands of DNA differences, each with tiny effects. Instead of hunting for big beasts in the genome jungle, we would be looking for microscopic creatures. This meant sample sizes not in the hundreds or even thousands but in the tens of thousands.

Even though I am an incorrigible optimist, a decade ago I was getting depressed about these three false starts and their implications for future attempts to find the DNA differences responsible for the heritability of psychological traits. I pondered retirement and changing my lifestyle. I contemplated a transatlantic sailing trip, thinking I might want to live permanently on a sailboat when I retired. Sailing across the North Sea in a warm-up voyage, I had a frightening experience one night, colliding with a barely submerged container the size of our sailboat that had broken loose from a cargo ship. I decided to stick with genetics and returned to my desk.

My misery about these false starts had lots of company, because many other GWA studies failed to come up with replicable results. The message slowly sank in that there are hardly any associations of large effect. The way out was to accept that much larger GWA studies would be needed to find the many tiny DNA differences responsible for heritability. At least this was beginning to look more feasible, because the price of SNP chips kept going down. Nonetheless, research funds with the huge sample sizes needed to detect such small effects would be available only for major medical disorders, not for psychological traits, especially controversial ones like intelligence.

One study pointed the way. In 2007, a GWA study was published that reported analyses of 2,000 cases for each of seven major disorders. These disorders included coronary artery disease, Type 2 diabetes and Crohn's disease, a chronic inflammatory bowel disease. Only one psychological disorder was included, bipolar disorder, which used to be called 'manic depression' because of its severe mood swings from mania to depression.

Most researchers had samples no larger than a few hundred cases. To reach the threshold of 2,000 cases for each of the seven disorders, researchers needed to pool their precious samples, often painstakingly collected over decades. This study led the way towards collaboration by bringing together over fifty UK research groups, with 258 co-authors on the 2007 paper. All 14,000 cases, as well as controls, were genotyped on a new SNP chip with half a million SNPs.

This visionary big-science study, funded with £10 million from the Wellcome Trust and a dozen other UK agencies, was called the Wellcome Trust Case Control Consortium. Across the seven disorders, twenty-four genome-wide significant SNP associations were found, mostly for Type 2 diabetes and Crohn's disease.

This Wellcome Trust study was a cause for celebration because it showed that GWA studies with large sample sizes could be successful even for common disorders influenced by many DNA differences of small effect. One index of the importance of this paper is that it has been cited more than 5,000 times in other scientific papers. In addition, GWA won the 'Breakthrough of the Year' in 2007 awarded by *Science*.

Despite the breakthroughs of the Wellcome Trust project, it was disappointing to see that 2,000 cases netted so few SNP associations and it was shocking to find that the effect sizes of the associations were all very small. As a psychologist, I was most disappointed that the only psychological disorder, bipolar disorder, showed no solid SNP associations.

The large expense of GWA studies and their low yield led to sniping about the cost–benefit ratio of GWA research, especially for psychological disorders. By 2011 the carping got so bad that ninety-six leading GWA researchers felt it necessary to publish a letter with the title 'Don't Give up on GWAS'. They concluded that failures were due to low power to detect small associations. GWA samples of sufficient size were being assembled that promised to be more successful.

The beacon of hope was the solid evidence that heritability is substantial. This means that inherited DNA sequence differences lurking in the genome make a big difference in psychological traits. So where were they? The most likely answer was that the effects of individual SNPs are even smaller than anyone expected. A sample of 2,000 cases, which seemed huge at the time, only had power to detect SNP associations that now seem unrealistically large.

For common disorders like bipolar disorder, with a prevalence of 1 per cent, a study with 2,000 cases could only detect a SNP association that increased the risk of having the disorder from 1 per cent to 1.6 per cent, a 60 per cent increase in risk. To find SNPs that increase risk by 30 per cent, samples with 10,000 cases would be needed. SNPs that increase risk by 10 per cent would need samples of 80,000 cases, which seemed ludicrously large for research on psychological disorders, where studies rarely included even a hundred cases, let alone thousands.

This new threshold of 80,000 cases motivated more researchers to collaborate, because they knew that their individual studies, usually with sample sizes of fewer than a thousand cases, had no power to detect associations of the size we now knew could be expected. In the biological and medical sciences more than a thousand GWA studies were reported in the five years following the Wellcome Trust study. Great progress was made during these five years, going from the twenty-four significant associations for seven traits from the Wellcome Trust study to more than 2,000 SNP associations for more than 200 traits. After five more years, in 2017, the number of genome-wide significant SNP associations had reached 10,000.

In psychology, a remarkable collaboration emerged, called the Psychiatric Genomics Consortium (PGC), which now includes over 800 researchers from more than 40 countries. The PGC focuses on the major psychological disorders other than Alzheimer's disease: schizophrenia, bipolar disorder, major depressive disorder, autism, hyperactivity, substance abuse, eating disorders, Tourette syndrome, obsessive–compulsive disorder and post-traumatic stress disorder.

Finding tens of thousands of cases is not as difficult as it might seem for psychological disorders because these disorders are, unfortunately, so common. For example, schizophrenia has a prevalence of 1 per cent, which means that in the UK alone more than half a million people suffer from it. The PGC has shown that bigger is better when it comes to GWA sample size. A 2014 report from the PGC for schizophrenia included 30,000 cases and netted more than a hundred genome-wide significant associations. By 2017 the PGC had doubled the number of cases and increased the catch to 155 associations.

For bipolar disorder, the PGC has gone from 2,000 cases in the Wellcome Trust study to 20,000 cases. The number of genome-wide significant hits has gone from zero to thirty. The PGC currently is working towards 50,000 cases.

Major depression got off to a slow start, with only one significant hit in a GWA analysis of 20,000 cases. In 2017 the PGC reported a GWA analysis of over 100,000 cases that identified 44 significant hits.

GWA studies of other psychological disorders are beginning to catch up with schizophrenia, bipolar disorder and major depression in sample size and in significant GWA results. For example, a recent

GWA study of hyperactivity with 20,000 cases reported 12 hits. The PGC is aiming for 40,000 cases each for hyperactivity, anorexia and autism. Most other psychological disorders, such as alcohol dependence and other substance-use disorders, anxiety disorders, post-traumatic stress disorder and obsessive–compulsive disorders are also targets of ongoing GWA studies.

What this means is that GWA hits are beginning to appear as studies of psychological disorders reach the power afforded by tens of thousands of cases. The results of GWA studies of psychological disorders confirm the daunting predictions from analyses of statistical power. With 10,000 cases, no significant associations are found. Significant associations begin to appear with 20,000 cases. Doubling the number of cases to 40,000 quadruples the number of significant hits. Doubling the sample size again to 80,000 shows another large increase in significant hits as power is reached to scoop up many of the smaller effects.

Like the Wellcome Trust results, the PGC results are a cause for celebration and for caution. They show that GWA is successful when sample sizes are sufficiently large. Finding 155 reliable associations for schizophrenia, 30 for bipolar disorder and 44 for major depression is a remarkable achievement. For the first time, we have reliably identified some of the DNA differences responsible for the heritability of psychological traits. It opens the door to the world of personal genomics, where we can use DNA differences across the genome to predict psychological differences between us. As we shall see, our passport to this new world was the ability to aggregate the effects of many tiny associations to predict psychological differences, or polygenic scores. For schizophrenia, DNA differences packaged as polygenic scores are now the best predictor we have of who will become schizophrenic. The rest of this book is about these polygenic scores and their impact on psychology and society.

One exception to the rule that there are no DNA differences that have a large effect on psychological traits is late-onset Alzheimer's disease. Although Alzheimer's disease is often regarded as a medical or neurological disorder rather than a psychological one, its early signs are purely psychological, especially memory loss regarding recent events. Alzheimer's disease typically afflicts people in their seventies and eighties. It accounts for more than half of all cases of dementia, and affects about 10 per cent of the population. Eventually, sometimes after fifteen years, individuals with Alzheimer's disease become bedridden, with extensive problems in brain nerve cells.

In 1993, a decade before the advent of GWA studies, a gene involved in cholesterol transport, apolipoprotein E (*APOE*), was found to be strongly associated with Alzheimer's disease. One of the *APOE* alleles, called allele 4, has a frequency of 40 per cent in individuals with Alzheimer's disease, as compared to 15 per cent in controls. Having two copies of allele 4 increases the risk of Alzheimer's disease from 10 per cent to 80 per cent. Fortunately, only 1 per cent of the population has two copies of allele 4. Half of individuals with Alzheimer's disease do not have any copies of allele 4, which means that allele 4 does not by itself cause Alzheimer's disease. A 2013 GWA analysis of Alzheimer's disease with 17,000 cases identified five other SNP associations of much smaller effect size that replicated in an independent sample of 8,000 cases.

For psychological disorders, more than a hundred GWA studies have been reported. Despite their large sample sizes, the biggest effects found in these first successful GWA studies of disorders, with the notable exception of Alzheimer's disease, were much smaller than anyone anticipated, only raising risk from 1 to 1.2 per cent. This is a 20 per cent relative increase in risk, but the absolute increase is only 0.2 per cent. Effects of this size are seen when the allele frequency for a SNP differs just slightly between cases and controls, for example, 45 per cent versus 40 per cent.

But if these tiny effects are the biggest effects scooped up by GWA studies with tens of thousands of cases, this means that most effects must be much smaller. With 80,000 cases, we can detect SNPs that add 10 per cent to the genetic risk for developing a disorder. But what if SNPs add only 1 per cent risk? Instead of 80,000 cases, millions of cases would be needed to detect such tiny effects. There are enough people in the world to find millions of individuals with schizophrenia, but it will be a challenge to find the money for such large GWA studies.

*

One way around this problem is to study dimensions rather than disorders. Dimensions provide more power in GWA studies than disorders because every individual counts, whether they are low, middle or high in the distribution. In contrast, GWA studies of disorders look for average DNA differences between two groups, cases who are diagnosed with the disorder versus controls who do not have the disorder. This assumes that disorders are real, but this assumption clashes with one of the big findings of genetic research – that the abnormal is normal, meaning that there are no qualitative disorders, just quantitative dimensions. The many DNA differences that are associated with what we call a disorder affect people throughout the distribution. GWA studies based on comparing diagnosed cases and controls lose a lot of information because many so-called controls will be close to being cases. This clouds the difference between cases and controls.

For example, SNPs associated with obesity are not SNPs for the diagnosis of obesity. They are associated with body mass index (BMI) throughout the distribution, from thin to heavy people, as we saw for the *FTO* SNP. In other words, these SNPs associated with BMI make a thin person a bit heavier just as much as they make an overweight person a bit heavier. We all have many of the SNP alleles that contribute to BMI. Being overweight is a matter of how many of these alleles you have. Obesity is not a qualitative disorder; it is a matter of more or less. This is what is meant by saying that complex disorders are quantitative traits, even for severe psychological disorders such as schizophrenia, bipolar disorder and autism. For this reason, polygenic scores will encourage psychology to move away from categorical diagnoses towards continuous dimensions assessed using standardized dimensional rating scales of symptoms, one of the important implications of the DNA revolution, as we shall see.

Another huge advantage of studying dimensions rather than disorders is that the same sample can be used to study many traits, whereas samples selected for a particular disorder are only useful to study that disorder. In many countries, biobanks have been set up with sample sizes in the hundreds of thousands that have collected a broad range of psychological as well as medical information. For example, UK Biobank, begun in 2006 and funded by UK charities and the British government, includes half a million volunteers who have provided DNA and access to their medical records and who have completed many measures, including measures of psychological dimensions. Similar projects are under way in other countries, including Estonia, the Netherlands and the Scandinavian countries. Finland has recently announced that it has begun a biobank that will obtain DNA from over 1 million people.

In the past two years there has been a surge of successful GWA studies of psychological dimensions. The first breakthrough was for an unlikely variable: years of education. In developed countries heritability of years of education is 50 per cent. Many psychological traits contribute to this heritability, such as previous achievement at school and cognitive abilities, which correlate 0.5 with years of education. The variable years of education is also affected by personality traits such as perseverance and conscientiousness, and mental health such as the absence of debilitating depression.

The reason for its GWA success is that a sample of more than a million people has been included, the largest GWA study to date. This large sample provided the power to pick up tiny SNP associations, which paid off in identifying more than a thousand genome-wide significant associations. Like all other complex traits, the effect sizes for years of education are incredibly small – the largest effect was only 0.03 per cent and the average effect size of the top SNPs was 0.02 per cent, which counts for just two weeks of education. However, as I will explain later, aggregating these SNPs can predict more than 10 per cent of the variance in years of education. This makes DNA the best predictor we have of a child's years of education, even better than the environmental effect of family socioeconomic status. This success signals the start of the DNA revolution in psychology.

GWA studies of other psychological dimensions have also been successful as their sample sizes became large enough to scoop up the many small SNP associations that are responsible for heritability. For intelligence, GWA studies had only modest success until sample sizes reached almost 300,000, when more than 200 significant associations were reported in 2018. Previous studies, including mine, did not have the power to detect these small associations.

Dozens of GWA studies have been reported for specific abilities such as reading and mathematics but their sample sizes have been too small to find many reliable associations. This situation will soon change as large-scale consortia are being created, for reading, for example, and some of the big biobanks are including measures of specific abilities.

Gene-hunting is also beginning to be successful for personality as the sample sizes of GWA studies of personality have increased. Because studies of personality rely on self-report questionnaires, it has been easier to obtain big samples for personality than for cognitive abilities, which require the administration of tests. The first successes have come in GWA studies of the two major dimensions of personality, extraversion and neuroticism, which twin studies indicate are about 40 per cent heritable. Extraversion includes sociability, impulsiveness and liveliness. Neuroticism, which refers to emotional instability rather than being neurotic, involves moodiness, anxiousness and irritability. For extraversion, a GWA study of 100,000 individuals found 5 hits. For neuroticism, over 100 hits were reported in a GWA study with a sample size of 300,000. A newer focus of personality research is a sense of well-being, basically happiness, which shows a similar heritability of 40 per cent in twin studies. In a GWA study of nearly 200,000 individuals, 3 hits were found.

GWA studies of other interesting personality-related traits are popping up. Many of these are from the UK Biobank, with its sample of half a million. Significant hits have been reported for traits such as coffee and tea consumption, chronic sleep disturbances (insomnia), tiredness, and even whether an individual is a morning person or a night person. Another recent example is a trait called cognitive empathy, which involves detecting emotions from photographs of eyes alone.

This is just the beginning of the DNA revolution. By the time you read this there will be dozens of bigger and better GWA studies of these and many other traits. An important source of new information will come from the biggest direct-to-consumer genomics company, 23andMe, with nearly 2 million paying customers. Eighty per cent of its customers have agreed to have their genotypes used in research and to consider follow-up requests for research. The average customer contributes to more than 200 brief studies, many of which are psychological studies.

The most shocking discovery from two decades of gene-hunting is that, instead of hunting for big game, the actual quarry are microscopic creatures. The effect sizes of the DNA differences responsible for the heritability of all psychological disorders and dimensions are much smaller than anyone anticipated. That is, twenty-five years ago, everyone hunting for genes assumed that a handful of genes accounted for most of the heritability observed in twin studies. As noted earlier, just ten genes that each accounted for 5 per cent of the variance would explain a heritability of 50 per cent.

The GWA results tell a very different story. For complex traits, no genes have been found that account for 5 per cent of the variance, not even 0.5 per cent of the variance. The average effect sizes are in the order of 0.01 per cent of the variance, which means that thousands of SNP associations will be needed to account for heritabilities of 50 per cent.

The brute force strategy of ever-larger samples to detect ever-smaller effects has paid off, so we now have thousands of SNPs associated with complex psychological traits. New refinements in the fast-moving science of genetics will increase the haul. One certain boost will come from genotyping all DNA differences, not just those currently on SNP chips. SNP chips used in GWA studies rely on common SNPs, those with allele frequencies greater than 1 per cent in the population, whereas the vast majority of DNA differences in the genome are much less frequent than 1 per cent. Many inherited DNA differences are unique to an individual.

These DNA differences can be genotyped with whole-genome sequencing that sequences all 3 billion base pairs of DNA. Whole-genome sequencing is the next big thing in genomics. It's the end of the story in the sense that the sequence of 3 billion base pairs of DNA is all that we inherit. This means that the inherited DNA differences responsible for heritability must be there somewhere.

Whole-genome sequences have already been obtained for hundreds of thousands of individuals. It has been predicted that in the next few years a billion individuals will have their whole genome sequenced and this DNA information will be linked to electronic medical records. We already know that there is an excess of rare mutations in individuals with schizophrenia, autism and intellectual disability and that individuals of extremely high intelligence have fewer of these rare mutations, suggesting that rare mutations are not good for you.

Without doubt, much will be learned from sequencing the 3 billion base pairs of DNA. It is a safe bet that, looking back a decade from now, we will realize how little we knew about how to find the DNA differences that make a difference in psychological traits. This new knowledge will increase our ability to find more of the inherited DNA differences that are responsible for the heritability of psychological traits.

Now that you have read this saga showing the need for larger and larger GWA studies to detect smaller and smaller SNP associations, it would be reasonable to ask, why bother? There are two reasons for hunting for the inherited DNA differences underlying individual differences in psychological traits. The first is to find pathways from genes to brain to behaviour and the other is to predict behaviour.

What good are such small effects? The answer is 'not much', if you are a molecular biologist wanting to study pathways from genes to brain to behaviour, or if you are in the pharmaceutical industry wanting to find a drug to fix a broken gene. Such small effects create a welter of minuscule paths that are difficult to track. Pinning down the mechanisms underlying SNP associations will be difficult because their effects are so small, about 0.01 per cent on average.

Further complicating this bottom-up pathways approach from DNA to behaviour is pleiotropy, which, as we have seen, means that any DNA difference affects many traits. Pleiotropy guarantees that there is no clear path from genes to brain to behaviour. The paths wander all over the brain. For example, the *FTO* SNP does not follow a straight path through the brain to affect our eating behaviour. Although the *FTO* gene is most well known for its effects on fat cells, it is highly expressed throughout the brain, especially in the cerebral cortex, which is centrally involved in all cognitive processes. These peripatetic effects are not special to the *FTO* gene – most genes affect most brain and behavioural processes. If each gene affects many behaviours, this means that each behaviour will be affected by many genes, which is exactly what the GWA studies have shown.

Another reason why a bottom-up approach from genes to

behaviour will be difficult is that most SNP associations with psychological traits do not involve genes in the traditional sense. The great majority of SNP associations have been found in non-coding regions of the genome. Little is known about this 'dark matter' of DNA – the 98 per cent of DNA that does not code for proteins. What we know so far is that non-coding regions can be involved in regulation of gene expression.

In contrast to the biologists' bottom-up-pathways game plan is the psychologists' top-down approach. For biologists, the ultimate goal of genetics is to understand every path between inherited DNA differences and individual differences in behavioural traits, a bottom-up approach. However, psychologists focus on behaviour and use genetics to understand behaviour. This top-down psychological perspective begins with prediction. We can use inherited DNA differences to predict individual differences in psychological traits without knowing anything about the myriad pathways connecting genes and behaviour.

The problem is that DNA differences that have such small effects seem worthless for prediction. A decade ago, as the realization sunk in that the biggest associations are extremely small, I had a thought that brightened the picture for me. Although the effects of individual SNPs are tiny, these effects can be added like we add items on a test to create a composite score. In 2005 I called these *SNP sets*. There are now at least a dozen names for these composite scores, but they are generally called *polygenic scores*.

Thinking about so many SNPs with such small effects was a big jump from where we started twenty-five years ago. We now know for certain that heritability is caused by thousands of associations of incredibly small effect. Nonetheless, aggregating these associations in polygenic scores that combine the effects of tens of thousands of SNPs makes it possible to predict psychological traits such as depression, schizophrenia and school achievement.

12

The DNA fortune teller

It has been known for decades that the heritability of psychological disorders and dimensions is caused by many DNA differences, not just one or two genes that pack a big punch. The shock from genome-wide association studies was the realization what 'many' meant – not a few dozen DNA differences but tens of thousands. GWA studies have shown that there are no associations that account for more than 1 per cent of the differences between individuals and that the average effect size is less than 0.01 per cent. This means that thousands of DNA differences contribute to the heritability of psychological traits and that huge GWA sample sizes are needed to detect these tiny associations.

After the false start of candidate-gene studies that failed to replicate, GWA research set a stringent criterion for reporting statistically significant 'hits' by correcting associations for a million tests across the genome. This criterion missed the many associations that do not, and cannot, reach statistical significance because their effects are so small. No matter how tiny these effects are, they can be combined to create a composite score, or polygenic score. Although the minuscule effects of individual SNPs are useless for prediction, polygenic scores that aggregate these effects, no matter how small, can powerfully predict genetic propensities. The 'poly' of 'polygenic' is what makes these scores able to predict individual differences in psychology. In other words, the key criterion for a GWA study is not how many associations reach statistical significance. Much more important is the power of a polygenic score derived from the results of a GWA study to predict individual differences.

Polygenic scores, based on DNA rather than crystal balls, are fortune tellers. As we shall see, prediction is crucial because it is the key to the prevention of psychological problems and the promotion of promise. This is the new world of personal genomics, which begins with the ability to use inherited DNA differences across the genome to predict psychological differences. For psychological dimensions and disorders, some polygenic scores have already reached impressive levels of predictive power. This chapter shows what a polygenic score is and describes the power of polygenic scores created in the past two years. It reveals some of my own polygenic scores to glimpse the future of psychological personal genomics.

Because polygenic scores are the basis for the DNA revolution in psychology, it is essential to understand what they are. A polygenic score is like any composite score that psychologists routinely use to create scales from items, such as those on a personality question-naire. The goal of a polygenic score is to provide a single genetic index to predict a trait, whether schizophrenia, well-being or intelli-gence. To get a concrete understanding of a polygenic score, consider a personality trait like shyness. A questionnaire to assess shyness includes multiple items in order to tap into different facets of shyness. For example, a typical shyness questionnaire will have items about how anxious you are in social situations and how much you avoid these situations – for example, going to a party, meeting strangers and speaking up at a meeting. You might be asked to respond using a three-point scale (0 = not at all, 1 = sometimes, 2 = a lot).

A shyness score is created by adding these items, taking care to 'reverse' items as needed so that a high score means a high degree of shyness. If our shyness measure had ten items scored 0, 1 and 2, total scores could vary from 0 to 20. Simply adding the items like this treats each item as if it is equally useful, but all items are not equally useful. For this reason, items are often added after they are weighted by some criterion of their usefulness at capturing the construct of shyness.

This is exactly how polygenic scores are created, except that, instead of items on a questionnaire, we add up SNP genotypes. Like the three-point rating scale for shyness, SNP genotypes are scored as 0, 1 or 2, indicating the number of 'increasing' alleles, as in the example of the *FTO* SNP. In the same way that we can add up alleles for one SNP to create a genotypic score, we can also add up alleles for many SNPs to create a polygenic score, just as we add questionnaire items to create a shyness score. The results from genome-wide association studies are used to select SNPs and to assign weights to each SNP. For example, in the GWA analysis of weight, the *FTO* SNP accounts for much more variance than other SNPs, so it should count for much more in a polygenic score for weight.

The following table shows how one individual's polygenic score is created from ten SNPs. For the first SNP, this individual's genotype is AT. For this SNP, the T allele happens to be the increasing allele that is positively associated with the trait. So, the individual's genotypic score for this SNP is 1 because the genotype has only one increasing T allele. Across the ten SNPs, the individual has a total of nine increasing alleles for the trait out of a possible score of 20. So, this individual would have a polygenic score just below the population average score of 10 for this trait.

This score merely adds the number of increasing alleles, which works reasonably well as a polygenic score. However, we can increase its precision by weighting the genotypic score for each SNP by how much

Table 3 A polygenic score for one individual based on ten SNPs

	Increasing allele	Allele 1	Allele 2	Genotypic score	Correlation with trait	Weighted genotypic score
SNP 1	T	A	T	1	.005	.005
SNP 2	C	G	G	0	.004	.000
SNP 3	A	A	A	2	.003	.006
SNP 4	G	C	G	1	.003	.003
SNP 5	G	C	C	0	.003	.000
SNP 6	T	A	T	1	.002	.002
SNP 7	C	C	G	1	.002	.002
SNP 8	A	A	A	2	.002	.004
SNP 9	A	T	T	0	.001	.000
SNP 10	C	C	G	1	.001	.001
Polygenic score				9		.023

the SNP correlates with the trait. The correlation between each SNP and the trait is taken from the GWA analysis. If one SNP correlates five times more with the trait than another SNP – such as SNP 1 versus SNP 10 – it should count for five times as much in the polygenic score.

The weighted genotypic scores in the last column of the table are the product of the genotypic score for each SNP and the correlation with the trait. The sum of these weighted genotypic scores for the ten SNPs is 0.023. This number isn't as interpretable as the unweighted genotypic score of 9, which is just the sum of the 'increasing' alleles. However, both the unweighted polygenic score of 9 and the weighted score of 0.023 can be expressed simply as a percentile in the population. For this individual, both types would indicate a polygenic score just below average.

How many SNPs should go into a polygenic score? Initially, polygenic scores were created using only the genome-wide significant 'hits' from a GWA study. For weight, ninety-seven independent SNPs reached genome-wide significance. Creating a polygenic score from these top ninety-seven SNPs explains 1.2 per cent of the variance in weight in independent samples. This is only slightly better than the prediction from the *FTO* SNP by itself, which explains 0.7 per cent of the variance.

Using only genome-wide significant hits is like demanding that each item in our shyness scale predicts significantly on its own. We don't do this for other psychological scores because it is unrealistic to expect each item to stand on its own. The goal is to have a composite scale that is as useful as possible.

A better idea is to do what we do when we create other psychological scores: keep adding items as long as they add to the reliability and validity of the composite in independent samples. For polygenic scores, the key criterion is prediction. The new approach to polygenic scores is to keep adding SNPs as long as they add to the predictive power of the polygenic score in independent samples. This is the strategy that has paid off in the last two years in producing powerful polygenic scores for psychological traits. Some false positives will be included in the polygenic score but that is acceptable as long as the signal increases relative to the noise, in the sense that the polygenic score predicts more variance.

For example, for BMI, a polygenic score based on the ninety-seven genome-wide significant SNPs predicts 1 per cent of the variance, but a polygenic score that includes 2,000 SNPs predicts 4 per cent of the variance of BMI. Including even more SNPs in the polygenic score increases the prediction to 6 per cent of the variance. Many false positive SNPs sneak into this polygenic score, but they don't hurt the prediction, they just don't help. Increasing the predictive power for the polygenic score from 1 per cent to 6 per cent makes this a very acceptable trade-off between signal and noise.

For complex traits and common disorders, this new approach to polygenic scores includes not just ten or a hundred or even a thousand SNPs. Typically, tens of thousands of SNPs are included in polygenic scores, sometimes hundreds of thousands. It's empirical – keep adding SNPs as long as they increase the power to predict in independent samples.

GWA summary statistics needed to create polygenic scores are currently available for hundreds of traits across biology and medicine, as well as psychology. After publishing a GWA study many researchers make their GWA summary statistics publicly available so that they can be used by anyone to create polygenic scores. To give a sense of the explosion of GWA research during the past decade, the main repository for these results includes GWA summary statistics for 173 traits based on 1.5 million individuals and 1.4 billion SNP-trait associations. These traits include twenty psychological traits and disorders and variables that are relevant to psychology, such as social deprivation, smoking, sleep duration, age at menarche and menopause and father's age of death. They also include physiological traits relevant to psychology, such as immunological and metabolic biomarkers.

It cannot be overemphasized that this is just the beginning of the era of polygenic scores. Although GWA summary statistics are publicly available for more than 200 traits, GWA analyses have been reported for hundreds of other traits that will eventually add to the list of possible polygenic scores as their summary statistics are made available. Also, bigger and better GWA studies will continually produce more powerful polygenic scores for all traits.

*

In the rest of this chapter, I will share my polygenic scores for height and weight in order to explore some general issues raised by these indicators. These provide concrete illustrations of how polygenic scores herald the era of personal genomics in psychology. In creating my polygenic scores, we used the most recently published GWA studies described in the previous chapter, although, in each case, GWA analyses of much larger samples are in the works. By the time you read this, the following account will be a conservative estimate of the power of polygenic scores.

Polygenic scores require DNA, genome-wide genotyping and lots of analysis. For less than £100, direct-to-consumer companies will extract your DNA from saliva and conduct genome-wide genotyping using SNP chips. These companies have focused on single-gene disorders, but the same genome-wide genotyping can be used to create polygenic scores. Companies are beginning to re-analyse SNP genotype data to provide polygenic scores for the general public. The same SNP genotypes can be used to create polygenic scores for any trait for which GWA results are available.

In order to illustrate polygenic prediction with my own DNA, a large comparison sample is needed with similarly constructed polygenic scores for each individual. Then we can see where my polygenic scores lie in the distribution for any of the hundreds of polygenic scores currently available. No phenotypic data are needed – just DNA. That is, without knowing anything about my depressive symptoms, I can compare my polygenic score for depression to the polygenic scores of the comparison sample.

My team and I created polygenic scores for a wide range of traits, using my SNP genotypes obtained from a SNP chip in our lab based on the results of the GWA studies described in the previous chapter. We compared my polygenic scores to those from a sample of 6,000 unrelated individuals participating in my UK-representative Twins Early Development Study. It doesn't matter that this comparison group comprises young adults because DNA does not change – the comparison group could just as well be infants.

The most predictive polygenic score so far is height, which explains 17 per cent of the variance in adult height. Although height is not a psychological trait, it is useful as a dispassionate example for understanding how polygenic scores work and how to interpret them. We used the GWA results for height to create polygenic scores for myself and for each individual in the TEDS sample. The polygenic score derived from the GWA of adult height predicts 15 per cent of the variance of height for the young adults in TEDS.

To interpret polygenic scores, it is important to keep in mind that they are always distributed like a bell-shaped curve, that is, normal distribution. This bell-shaped curve is dictated by the fundamental law of probability, or the *central limit theorem*, which is the basis for all statistics. The normal distribution is found when many random events contribute to a phenomenon, like flipping a coin and counting the number of times the coin comes up heads. If you flip a coin ten times, you could get no heads or ten heads in a row, but most of the time the total number of heads will be between four and seven. If you do this many times, you will get a perfectly normal bell-shaped distribution, peaking at five, which will be the average number of heads. Flipping coins and counting heads is exactly analogous to counting the numbers of 'increasing' alleles from SNPs to construct polygenic scores for many individuals.

I will describe all my polygenic scores in terms of percentiles in the normal distribution. That is, to what extent is my polygenic score above or below the average polygenic score in the comparison sample, the 50th percentile? It turns out that my polygenic score for height is at the 90th percentile. So, based on my DNA alone, knowing nothing else about me, you could predict that I am tall. And, in fact, I am 6 feet 5 inches. Of course, you can easily see that I am tall if you saw me, but with DNA you could tell that I am tall without even looking at me.

Most importantly, you could have predicted when I was born that I would be tall. Unlike any other predictors, polygenic scores are just as predictive from birth as from any other age because inherited DNA sequence does not change during life. In contrast, height at birth scarcely predicts adult height. The predictive power of polygenic scores is greater than any other predictors, even the height of the individuals' parents. Another advantage of polygenic scores over family resemblance is that parental height provides only a family-wide prediction that is the same for any child born to those parents. In contrast, polygenic scores provide a prediction specific to each individual. In other words, my polygenic scores at birth would have predicted that I would be taller than expected on the basis of the average height of my parents.

Before looking at my other polygenic scores, one other general point needs to be highlighted about predicting individuals. My actual height is at the 99th percentile but my polygenic score is at the 90th percentile. Are polygenic scores sufficiently accurate for prediction?

For example, in TEDS, the polygenic score for height predicts 15 per cent of the variance in actual height in these young adults. But 15 per cent is a long way from 100 per cent. In fact, polygenic scores can never predict 100 per cent of the variance of any trait, because the ceiling for prediction is heritability. For height, heritability is 80 per cent, but for psychological traits heritability is 50 per cent, which means that polygenic score prediction is always going to be way south of perfect. The big question is the extent to which polygenic scores will be able to predict all the heritable variance of traits. This gap is called *missing heritability*, and is described in the Notes section at the end of this book.

The correlation between the polygenic score and height is 0.39 for the individuals in the comparison sample. Squaring a correlation tells us how much variance in height is explained by the polygenic score, which is where the estimate of 15 per cent comes from. Figure 5

Figure 4 My polygenic score for height

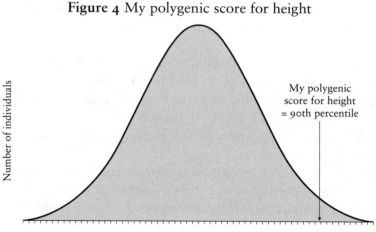

My polygenic
score for height
= 90th percentile

Number of individuals

Polygenic scores for heights in TEDS

Figure 5 Scatterplot showing the correlation of 0.39 between each individual's polygenic score for height and actual height, with my data point marked

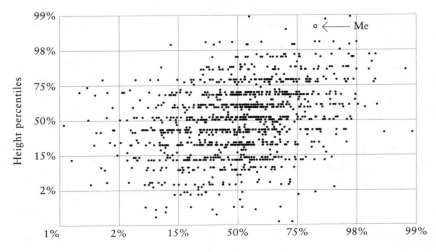

Note: Because of the large gender difference for height, height was corrected for gender and the results were standardized. For this reason, the results are presented as percentiles rather than in centimetres.

shows what the correlation of 0.39 looks like when you plot each individual's polygenic score against their height for the comparison sample.

If the correlation were 0, the scatterplot would look round rather than oval, indicating no association between polygenic scores and height. If the correlation were 1, the scatterplot would be a straight line. The prediction of height from the polygenic score for height is in between no prediction and a perfect prediction, as indicated by the correlation of 0.39.

You can see from the oval-looking scatterplot that higher polygenic scores are correlated with greater height. But there is variability. For example, my actual height is at the 99th percentile but my polygenic score is at the 90th percentile. Perhaps this discrepancy is due to some environmental factors, such as good nutrition or the absence of disease. But, more likely, it is just random fluctuation, given the moderate predictive power of the polygenic score.

There are much more extreme outliers than me. The highest polygenic score for height, on the far right of the figure, is for an individual whose actual height is slightly below average. At the other end of the distribution, the lowest polygenic score is for an individual whose actual height is near the population average.

Some scientists have used this inaccuracy to argue that polygenic scores cannot be used for individual prediction. The correlation between polygenic scores and height is not 1, and it cannot be 1 because heritability is less than 100 per cent and heritability is the ceiling for polygenic score prediction. However, the correlation of 0.39, explaining 15 per cent of the variance, gives us more predictive power than we have for other predictors, for example, predicting the height of individuals from their parents' height.

For any polygenic score, especially powerful predictions can be found at the extremes. For example, look at the scatterplot for height in Figure 5. You can see that the average height of individuals with low polygenic scores is much lower than the average height of individuals with high polygenic scores. Figure 6 makes this explicit by dividing the sample into ten equal-sized groups (deciles, each accounting for 10 per cent of the sample) on the basis of their polygenic scores for height and then calculating the average height of each group.

There is a strong relationship between average polygenic score and average height. For example, the average height of individuals in the lowest decile of polygenic scores is at the 28th percentile, whereas the average height of individuals in the highest decile of polygenic scores is at the 77th percentile.

The line running through each data point is called the standard error. The length of the line indicates the range of estimates that would be expected 95 per cent of the time. Note that the standard error refers to the average of each group, not the error of estimating an individual's score. In other words, the standard error surrounding the top decile means that 95 per cent of the time the *average* height of individuals in that decile would be between the 72nd and the 82nd

Figure 6 The average height of individuals from the bottom 10 per cent to the top 10 per cent of polygenic scores for height

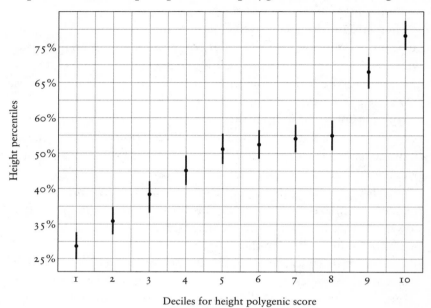

Deciles for height polygenic score

Note: Because of the large gender difference for height, height was corrected for gender and the results were standardized. For this reason, the results are presented as percentiles rather than in centimetres. The dots indicate the average height of individuals in each of the deciles for the polygenic score. The line running through each dot is the standard error of the average, which indicates the range of estimates that would be expected 95 per cent of the time.

percentile. It does not mean that the actual height of 95 per cent *of individuals* in the top decile of polygenic scores will be in this range.

The clearest way to express this crucial distinction between group differences and individual differences is to compare the distribution of scores for individuals in the groups with the lowest and highest polygenic scores. Figure 6 shows a big difference in the average height between the lowest and highest deciles of polygenic scores. Figure 7 shows the same mean differences in height but, in addition, it shows the distribution of individual differences around these group averages.

Figure 7 The distribution of height for individuals in the lowest and highest deciles of polygenic scores for height

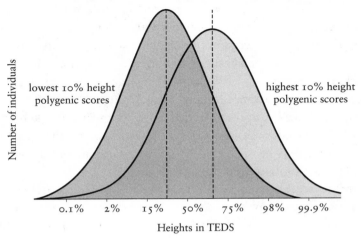

Despite the average height difference shown by the dotted lines, individuals within the two groups vary widely in height. The overlap between the two groups is 52 per cent, meaning that the group with the highest polygenic scores includes individuals shorter than most people in the group with the lowest polygenic scores, and vice versa.

So, if all you know about people is their DNA, you can predict their height. For groups of people with low or high polygenic scores, you can accurately predict that they will, on average, differ in height. However, when it comes to predicting the height of a single individual – you, for instance – prediction is less precise. Polygenic scores are useful for individual prediction only as long as we keep in mind that the prediction is probabilistic, not a certainty.

The ability of polygenic scores to predict height from birth might satisfy the curiosity of parents and help basketball scouts, but height does not have as much significance medically or socially as other traits. Weight, on the other hand, is correlated with many health outcomes and is a key variable in health psychology. Because of the strong correlation between height and weight, about 0.6, a purer measure of weight is used, *body mass index* (BMI), which corrects weight for height. For example, I weigh 114 kilograms (250 pounds) – corrected for height, gender and age, my BMI is thirty, which is at the 70th percentile for UK males of my age, whereas my actual height is at the 99th percentile.

I was shocked to find that my polygenic score for BMI is at the 94th percentile (Figure 8). My first thought was that this is an example of the lack of precision of polygenic scores, because my actual weight is at the 70th percentile. After all, the polygenic score for BMI predicts only 6 per cent of the variance, which is much less than the 15 per cent of the variance in height predicted by the polygenic score for height. However, upon reflection, my score seems unlikely to be a statistical fluke because it is so high. I also realized that my family tree has some very heavy limbs. Moreover, truth be told, I constantly struggle to keep my weight down.

I came to accept that my high BMI polygenic score makes sense. At any rate, accepting my BMI polygenic score has had a good effect on my attempts to persevere with my never-ending battle of the bulge, which serves as an example of how polygenic scores can enlighten self-understanding. The main point is that my high polygenic score does not mean that I must resign myself to being overweight. It means that I am genetically predisposed to put on the pounds and that I find it harder to lose them. Forewarned can be forearmed.

This genetic predisposition includes psychological as well as physiological mechanisms, such as sensitivity to food cues and the sense of

Figure 8 My polygenic score for body mass index (BMI)

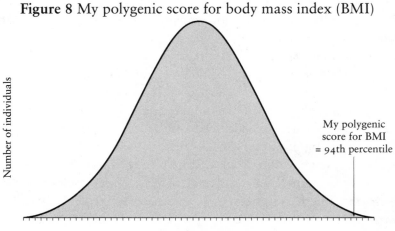

My polygenic score for BMI = 94th percentile

Number of individuals

Polygenic scores for BMI in TEDS

satiety. Knowing my BMI polygenic score helps me realize that I can't let my guard down, because it is in those weak moments – for example, when I am tired after a long day – that I sometimes give in to those siren snacks in the cupboard whispering to me. I know I am much better off if I just don't have any snacks available to tempt me. I can also see that I suffer from deficient satiety, meaning I struggle to stop eating even when I know I am full. Even after I know I am stuffed, I find it hard to resist finishing everything edible on the table. Simply being aware of my satiety deficit helps me to curb my overeating.

Self-understanding can also be enhanced by considering the discrepancy between polygenic scores and actual scores. Although my BMI polygenic score is at the 94th percentile, my actual BMI is 'only' at the 70th percentile. This discrepancy between my polygenic score and my actual BMI motivates me not to give up.

One general message that we should take from genetics is tolerance for others and for ourselves. Rather than blaming people for being overweight, we should recognize and respect the huge impact of genetics on individual differences. Genetics, not lack of willpower, is the major reason why people differ in BMI. Success and failure, credit and blame, in overcoming problems should be calibrated relative to genetic strengths and weaknesses.

Now that we've examined how polygenic scores for height and weight can foretell our future, let's turn to what polygenic scores can tell us about psychological traits. If you are interested in polygenic scores for common medical disorders, please see the Notes section.

13

Predicting who we are

Given that the focus of this book is psychology, the crucial question of course is what polygenic scores can reveal about psychological traits. After twenty years of trying unsuccessfully to find some of the inherited DNA sequences responsible for the substantial heritability of psychological traits, the last two years have been tremendously exciting. Creating polygenic scores using tens of thousands of SNPs has turned the tide in terms of predicting psychological traits. More powerful polygenic scores are pouring in every month.

In this chapter we will examine some of the best polygenic scores in psychology and see my own scores for these traits. Let's begin by looking at polygenic scores for the major psychological disorders of schizophrenia, depressive disorder and bipolar disorder.

For schizophrenia, polygenic scores can currently predict 7 per cent of the variance of the liability to be diagnosed as schizophrenic. (To learn more about what 'variance of the liability' means, please see the Notes section.) This 7 per cent is a long way from the 50 per cent heritability of schizophrenia, but it already predicts more of the liability variance than variables traditionally used to predict risk for schizophrenia, such as social disadvantage, cannabis use and childhood trauma such as bullying. Moreover, these 'environmental' correlations have not been controlled for genetics and are to some extent inflated. The polygenic score even predicts as well as family history, that is, knowing that a parent or sibling has been diagnosed as schizophrenic, which of course also includes genetic influence. Instead of the population risk of 1 per cent, having a first-degree relative that has suffered schizophrenia increases risk to 9 per cent. Conversely, this means that, more than 90 per cent of the time, individuals with a first-degree relative diagnosed as schizophrenic will not themselves be diagnosed as schizophrenic. In contrast, individuals who have the highest 10 per cent polygenic scores for schizophrenia are fifteen times more likely to be diagnosed as schizophrenic as the lowest 10 per cent. Moreover, ongoing GWA analyses have doubled the sample size, which will produce a substantial jump in the predictive power of its polygenic score.

As compared to schizophrenia, current polygenic scores for major depressive disorder and bipolar disorder predict less liability variance – 1 per cent for major depressive disorder and 3 per cent for bipolar disorder. However, these polygenic scores were based on about 10,000 cases each. GWA studies are in progress that greatly increase these sample sizes, which will add substantially to the predictive power of the polygenic scores. For major depressive disorder, the sample size has increased eightfold and the predictive power of the polygenic score has reportedly increased from 1 per cent to 4 per cent, although these analyses are still in progress. Four per cent is more predictive power than provided by traditional variables used to predict depression, most notably depression in parents. For bipolar disorder, doubling the sample size has increased the predictive power of the polygenic score from 3 per cent to 10 per cent in preliminary analyses, which is again the most powerful predictor we have for bipolar disorder.

Polygenic scores are also currently available for developmental disorders such as anorexia, autistic spectrum disorder and attention deficit hyperactivity disorder. However, so far, these polygenic scores account for less than 1 per cent of the variance of liability, which is not surprising, because each of the GWA studies from which they were derived included only about 3,000 cases. The predictive power of these polygenic scores will increase dramatically as current plans are realized to conduct GWA analyses for each of these disorders with sample sizes of 40,000 cases, more than ten times larger than the current GWA studies.

So, where do my polygenic scores for these psychological disorders fall?

I was most surprised by my polygenic score for schizophrenia – it was at the 85th percentile. I don't feel at all schizophrenic, in the sense of having disorganized thoughts, hallucinations, delusions or paranoia. Also, I don't know of any schizophrenia hiding in the branches of my family tree, including my son, who is forty years old and therefore past the usual age of onset.

If my higher than average polygenic score is not a statistical fluke and I am in fact genetically prone to schizophrenia, I can take some satisfaction in realizing that I have not succumbed. However, I wonder if my need for a highly structured, scheduled working life may be an attempt to keep myself on an even keel. One thing for sure is that this information makes me even less willing than I would normally be to try the new high-THC forms of cannabis that have been linked to onset of schizophrenia. On the other hand, I am well past the age of onset for schizophrenia, so I won't lose any sleep over my high polygenic score.

This is an example of the larger dilemma of what to do if we find out that we have a hefty genetic risk for a disorder that we can't do much about. For some problems, it is useful to know if you are at high risk because there are things you can do to lessen your risk. A good example is learning that I am at high genetic risk for being over-weight. Obviously, there are things I can do about that.

However, there are some psychological problems that you can't do much about at present, such as finding out that you have a high genetic risk for schizophrenia. Worse, what if your child has a high genetic risk for schizophrenia? As yet, there is little we can do to prevent these problems, other than common-sense things like avoiding mind-altering drugs. People differ in their reactions to this dilemma about discovering genetic risks when there is not much that can be done to fix the problem. To know or not to know, that is the question. Many people prefer not to know. Some, like me, prefer to know what may be in store for them, even if there is not much that can be done about it. Much has been written about the to-know-or-not-to-know question, although almost all of this is about single-gene disorders with their definitive answers about risk. Polygenic scores will always be probabilistic, not deterministic, because their ceiling is heritability, which is usually about 50 per cent. The closest that a genetic risk for a psychological trait gets to the concerns of a single-gene disorder is risk for Alzheimer's disease, which I will consider shortly.

Given the absence of any history of schizophrenic-like behaviour anywhere in my family, most likely I am a one-off, just getting a chance combination of SNPs that predispose to schizophrenia. In other words, my polygenic score may be the luck of the draw at conception, because genetic risk involves thousands of tiny DNA differences. This is why most people diagnosed with schizophrenia do not have any relatives who are schizophrenic, even though schizophrenia is substantially heritable. This is also the reason why polygenic scores are so important. Polygenic scores go beyond average family risk to predict genetic risk for each individual.

A nicer way of thinking about my higher than average polygenic risk score for schizophrenia is to contemplate possible positive aspects of what at the extreme is called schizophrenia. The best example is a possible link between schizophrenia and creative thinking. Aristotle said, 'No great genius was without a mixture of insanity.' Many artists have suffered from schizophrenia, most famously the painter Vincent Van Gogh, as well the novelist Jack Kerouac and the musicians Syd Barrett of Pink Floyd and Brian Wilson of the Beach Boys. Some especially creative scientists have also been diagnosed as schizophrenic, such as the mathematician John Nash, whose life was dramatized in the Hollywood movie *A Beautiful Mind*.

Recently, a family study of more than a million psychiatric patients in Sweden provided evidence to support these anecdotes, finding that the non-diagnosed first-degree relatives of schizophrenics were more likely to be in creative professions, such as actors, musicians and writers. A good example of the future use of polygenic scores is a recent study that asked whether the polygenic score for schizophrenia could predict creativity in healthy people. In several diverse populations the researchers found that people with high polygenic scores for schizophrenia were more likely to be in creative professions.

These thoughts will not be of much comfort to a parent who finds that their child has a very high polygenic score for schizophrenia. It is worth reiterating the mantra that polygenic scores are inherently probabilistic, not deterministic. Also, the ability of polygenic scores to predict problems makes it possible for research to focus on interventions that may eventually prevent or at least ameliorate these problems. We will return to these issues shortly.

For major depressive disorder and bipolar disorder, my polygenic scores were, respectively, at the 33rd and 22nd percentiles, suggesting a low risk. I was initially pleased with my lower than average polygenic scores for these major psychological disorders. However, we don't really know what low polygenic scores mean because psychologists have focused on diagnosed cases at the high end of the distribution. For example, my low polygenic score for bipolar disorder might do more than put me at low risk for experiencing the up-and-down mood swings of bipolar disorder. It might make me flat in affect, not smelling the roses. Failing to experience the highs and lows of life might also make me seem less empathic. It might even make me appear autistic. We have much to learn about the 'other end' of the distribution of polygenic scores. We will come back to this important implication of polygenic scores in the next chapter.

Because I am past the usual age of onset for these disorders, I wasn't worried about them as I waited for the results. However, late-onset Alzheimer's disease is a different matter altogether. The best that can be said for this horrible disease, described earlier, is that you have to live most of a long life before it gets you. A polygenic score for Alzheimer's disease can predict 5 per cent of the liability. Unlike other psychological disorders, most of this genetic risk is due to a single gene called *APOE*. Although only 1 per cent of the population has two copies of the recessive risk allele, the risk for these unlucky people jumps from the population risk of 10 per cent to 80 per cent, which is why this allele accounts for most of the predictive power of the polygenic score.

The effect of the *APOE* gene is big enough and Alzheimer's disease is scary enough that many people choose not to find out about their *APOE* status when they have their genome genotyped, including the first person to have his whole genome sequenced, James Watson, who shows no signs of Alzheimer's disease at ninety years of age. If there were something that could be done to prevent the downward spiral of this degenerative disease, people would be clamouring to get their polygenic score for Alzheimer's disease. The dilemma is about discovering genetic risks when there is nothing, as yet, that can be done about it.

I had to ask myself what I would do if I found out that I had two copies of *APOE* allele 4. Would it be better not to know, given that there is currently, in any case, no way to ward off this awful disease? I decided that I would, on balance, prefer to know – the knowledge-is-power argument. Finding that I had a substantial genetic risk for Alzheimer's disease would definitely make me plan my life differently. On the practical side, I would plan for care arrangements later in life. I would keep an eye on ongoing treatment trials and I would keep my fingers crossed for new treatments. Otherwise, the usual advice might help, for example, keeping my blood pressure under control, eating healthily and keeping active physically, mentally and socially. At least doing these things won't do any harm. The only specific advice would be to avoid head injury – definitely no boxing and probably no heading footballs – because head injury is the one environmental factor known to increase risk for Alzheimer's disease. Knowing that I was at high risk for Alzheimer's disease might also have some positive aspects, such as encouraging me to live more in the moment.

So I bit the bullet and looked at my *APOE* results. With great relief, I found that neither of my two alleles for *APOE* is allele 4. I am not especially lucky in this, because only 1 per cent of the population has two copies of *APOE* allele 4. Although more than a quarter of the population has one copy of allele 4, having a single copy harbours much less genetic risk for Alzheimer's disease. Because *APOE* does most of the heavy lifting for the polygenic score for Alzheimer's disease, my polygenic score is also lower than average, at the 39th percentile.

The biggest GWA study reported so far in all of science is for years of education, with a sample size of over a million. The huge sample size made it possible to uncover more than a thousand significant SNP associations. A polygenic score based on this study predicts more than 10 per cent of the variance in years of education, referred to as educational attainment. Although this new educational attainment polygenic score is not yet available, a polygenic score based on a GWA study with 330,000 individuals published in 2016 has taken psychology by storm, as we shall see, with dozens of papers already published, even though it predicts only 3 per cent of the variance in years of education.

What is my 2016 polygenic score for educational attainment? It turns out that this is my highest polygenic score, at the 94th percentile. This was welcome news, of course, but it led to some self-reflection. I grew up in a one-bedroom flat in inner-city Chicago without books. No one in my family went to university, including my parents, my sister and a dozen cousins who lived nearby. However, I was an avid reader from an early age, bringing bags of books home from my local public library. I often wondered where my interest in books and school came from, given that my family showed little interest in these things – for a while as an adolescent, I wondered if I had been adopted. I didn't realize then that, although the first law of genetics is that like begets like, the second law is that like does not beget like. Genetics makes first-degree relatives 50 per cent different as well as 50 per cent similar.

Although I always did well at school, I didn't think I was especially smart. I worked hard. I was conscientious. I persevered. I wonder if my high educational attainment polygenic score comes from the fact that the GWA target trait of years of education taps into a mishmash of traits needed to succeed in higher education, including interest in reading and personality traits such as conscientiousness and grit, in addition to intelligence. Research described later supports this hypothesis.

What if you found out that one of your children has a low score for educational attainment, which is quite possible, regardless of how high your polygenic score is? Even knowing that this is just a probabilistic prediction, it's a tough thing to accept, especially for highly educated parents. On the one hand, as emphasized repeatedly in this book, genes are not destiny and heritability describes what is, not what could be. Parents *can* make a difference. It is important that parents are not fatalistic about their children, because polygenic scores are probabilistic not deterministic.

On the other hand, as discussed earlier, it is also important that parents realize that children are not blobs of clay to be moulded however they wish. The main message of *Blueprint* is that genes are the major systematic force in children's development. Parents naturally want their children to be the best they can be, but it is important to distinguish that from what parents want their children to be.

Polygenic scores might help parents understand that a child's lack of interest in higher education is not necessarily a sign of recalcitrance or laziness. Learning is more difficult and less enjoyable for some children than others. In particular, polygenic scores could help parents who have more than one child to understand why one of their children takes to education but another does not.

The impact of the educational attainment polygenic score will rocket when the new polygenic score based on over a million individuals is available. Although years of education is a coarse measure, it is the best variable we have for predicting important social outcomes, most notably occupational status and income. Much of its predictive power derives from its correlation of 0.5 with intelligence. A surprising finding from research using the 2016 educational attainment polygenic score is that it predicts intelligence better (4 per cent) than it predicts its GWA target trait of years of education (3 per cent). The reason for this finding is that intelligence is assessed in a more refined manner.

A related curious finding is that it also predicts intelligence better (4 per cent) than do polygenic scores derived from GWA studies of intelligence itself (3 per cent). The reason for this is that the sample size for the GWA is larger and thus more powerful. The forthcoming educational attainment polygenic score based on a GWA sample of a million predicts more than 10 per cent of the variance in intelligence. It will be difficult for GWA studies of intelligence itself to reach similar sample sizes because intelligence has to be tested, whereas years of education can be assessed with a single self-reported item. Until much larger GWA studies of intelligence are conducted, this polygenic score will continue to be the best predictor of intelligence.

Because of my interest in school achievement, I wanted to see how well the educational attainment polygenic score predicts actual school achievement assessed by test scores, not just total years of education. No GWA studies have as yet focused on school achievement, so no polygenic scores are available to predict school achievement. In my UK TEDS twin study, we correlated the educational attainment polygenic score with test scores on the UK national examination given at the age of sixteen, GCSEs.

We found that the polygenic score created from the results of the

2016 GWA study of total years of schooling in adults predicts 9 per cent of the variance of GCSE scores at the age of sixteen. What this means is that the GWA analysis of years of education inadvertently did a better job at capturing genetic variation for actual school achievement (9 per cent) than it did for the target variable of years of education (3 per cent). In addition, using an approach called *multi-polygenic scores*, we were able to boost this result to predict 11 per cent of the variance in GCSE scores by including polygenic scores for intelligence in addition to the educational attainment polygenic score. Predicting 11 per cent of the variance makes it the strongest polygenic score prediction of any psychological trait reported as of 2017, although this record will soon be broken as results keep pouring in.

Few variables can predict school achievement this well. We have seen that the intensive and expensive on-site evaluations of school quality in the UK predict less than 2 per cent of the variance in children's GCSE scores at the age of sixteen. One of the best long-term predictors of children's school achievement is their parents' educational attainment. In TEDS, parental educational attainment predicts 20 per cent of the variance in their children's GCSE scores. However, we have shown that half of this correlation between parental educational attainment and children's GCSE scores is due to genetics, another example of the nature-of-nurture phenomenon. In other words, parental educational attainment predicts 10 per cent of the variance in GCSE scores, once we control for genetics. So, predicting 11 per cent of the variance from DNA alone is impressive.

As we saw for height, especially powerful predictions can be made at the group level from the educational attainment polygenic score. Figure 9 shows the strong relationship between these and GCSE scores when the polygenic scores of the TEDS sample are divided into ten deciles. The figure shows that the average GCSE score increases steadily as the educational attainment polygenic score increases. The real-world impact of polygenic scores can be observed at the extremes. Children in the lowest and highest educational attainment deciles differ by a full GCSE grade on average. Only 32 per cent of students in the lowest decile go to university, whereas 70 per cent in the highest decile go to university.

Figure 9 The average GCSE scores of individuals with increasing polygenic scores for educational attainment (EA)

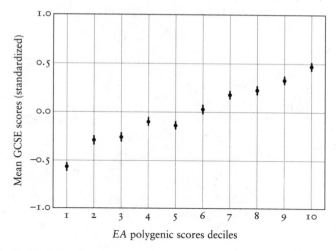

EA polygenic scores deciles

Note: The dots indicate the average GCSE score of individuals in each of the ten deciles from low to high EA polygenic scores. The line running through each dot is the standard error of the average, which indicates the range of estimates that would be expected 95 per cent of the time.

Despite educational attainment's strong prediction of group differences, prediction of individual differences is not precise. Although we explored this issue earlier in relation to height, the distinction between predicting group differences versus individual differences is so important that it is worth making the point again in relation to school achievement. Figure 10 shows the average difference in GCSE scores between the bottom and top deciles but adds the distribution of individual differences around these group averages.

The two groups differ substantially in their average GCSE scores, as shown by the dotted lines, which reiterates the difference shown in Figure 9. However, individuals within the two groups vary widely in their GCSE scores. The overlap between the two groups is 57 per cent. You can see that many individuals from the group with the lowest polygenic scores have higher GCSE scores than people in the group with the highest polygenic scores. And vice versa.

Figure 10 The distribution of GCSE scores for individuals with the lowest 10 per cent and highest 10 per cent polygenic scores for educational attainment (EA)

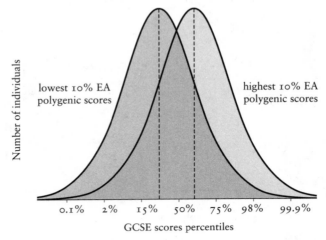

This result serves as another reminder that polygenic scores are only probabilistic predictors, as is the case for all the predictors that we use in psychology. This simply means that correlations are less than 1. Polygenic scores predict the mean outcome of groups quite well, such as groups with low versus high polygenic scores, but there is a wide range of individual differences within each group.

So, if all you know about people is their DNA, you can indeed predict their school achievement. The educational attainment polygenic score is already among the most powerful predictors in psychology. On the other hand, all polygenic scores are only probabilistic predictors and we need to remember that there is a wide range of individual differences in the target trait for each level of polygenic scores.

Not only does this polygenic score predict intelligence and tested school achievement, it predicts many other psychological traits, including personality and mental health. The reason for this is that many psychological traits are involved in educational attainment, not just intelligence and previous school achievement. For example, conscientiousness makes it more likely that a student will persevere, despite the stress and the ups and downs of further education. As Thomas Edison said, genius is 1 per cent inspiration and 99 per cent perspiration. Emotional stability helps too. Because educational attainment depends on several psychological traits, it is not surprising that educational attainment predicts many psychological traits, a second reason why this polygenic score is taking psychology by storm.

We have explored five of the best polygenic scores in psychology at the moment – schizophrenia, bipolar disorder, major depressive disorder, Alzheimer's disease and educational attainment. Polygenic profiles across psychological traits will paint a picture of an individual's genetic strengths and weaknesses. This has not been done previously, so Figure 11 summarizes my results as the world's first polygenic score profile for psychological traits. It shows my high polygenic scores for schizophrenia and educational attainment and my lower than average scores for bipolar disorder, major depressive disorder and Alzheimer's disease. Some of my other psychological polygenic scores are middling, for example, neuroticism, which is at the 66th percentile, and hyperactivity, which is at the 70th percentile, but I did not consider these polygenic scores strong enough at the present time to include them in my profile.

Polygenic profiles can include many more psychological traits than these five forerunners. It will soon be possible to extend profiles to another dozen psychological traits, including developmental disorders such as anorexia, autism and attention deficit hyperactivity disorder, specific cognitive abilities such as verbal and memory abilities, personality traits such as extraversion and well-being, and other traits such as sleep quality and whether or not you are a morning person. However, the polygenic scores included in Figure 11 are the polygenic scores most likely to be encountered in psychology in the next few years, because these traits have benefited from the largest GWA studies. A polygenic score for intelligence was not included in my profile because intelligence is currently predicted better by the educational attainment polygenic score. Polygenic scores for personality traits were not included because, so far, they do not explain much more than 1 per cent of the variance. For all these traits, we could boost the predictive power of polygenic scores using a multi-polygenic approach, as mentioned earlier, but for simplicity I chose to focus on single polygenic scores.

Figure 11 My psychological polygenic score profile

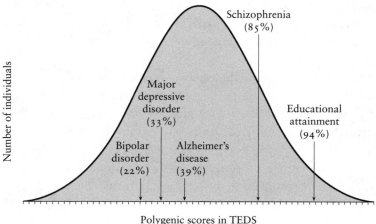

Polygenic scores in TEDS

Despite these caveats, these five forerunners of the DNA revolution in psychology will serve as examples of polygenic scores. By themselves, polygenic scores cannot yet be used to diagnose disorders, although they are already the best predictors we have for schizophrenia. Polygenic scores are also the best predictors of how well children will do at school.

It is important to remember that these are very early days in research on polygenic scores. It is a safe bet that the predictive power of most polygenic scores will double in the next few years. Because heritability is about 50 per cent, polygenic scores still have a lot of headroom for improving their predictive power. The reason for my obsession with predictive power is straightforward. The greater the predictive power of polygenic scores, the more valuable they will be for psychology and for society. This is the topic we turn to next.

14

Our future is DNA

Blueprint began with a sales pitch about a new fortune-telling device that promises to transform our understanding of ourselves and our life trajectories. It predicts important traits like schizophrenia and school achievement better than anything else, including family background, parenting and brain scans. It is 100 per cent reliable and 100 per cent stable, unchanging from day to day, year to year, birth to death, which means that it predicts adult traits from conception or birth just as well as it does in adulthood. The device is also unbiased, and not subject to coaching, faking or anxiety. And the one-time total cost for this new device is about £100.

I hope this no longer sounds like just another pop-psychology claim without evidence to back it up. The device is, of course, polygenic scores, backed up by the best science of our times.

Polygenic scores are the ultimate psychological test because, for the first time, they can tell our genetic fortunes. Although polygenic scores only tell us about genetic propensities, not about environmental effects, we have seen that inherited DNA differences are the major systematic cause of who we are. DNA differences account for half of the variance of psychological traits. The rest of the variance is environmental, but that portion of the variance is mostly random, which means we can't predict it or do much about it.

Even though polygenic scores have just burst on to the stage in the last few years, they are already beginning to transform clinical psychology and psychology research. As we enter the era of personal genomics, they will eventually affect all of us.

The transformative power of polygenic scores comes from three unique qualities. The first is that predictions from polygenic scores to psychological traits are causal, meaning that DNA differences cause differences in psychological traits. Predictions from polygenic scores are an exception to the rule that correlations do not imply causation. Earlier, we considered examples in which 'environmental' measures are assumed to be the cause of correlations with psychological traits – for example, correlations between parents reading to children and children's reading ability, between bad peers and bad adolescent outcomes, and between stress and depression. Always in psychology it is possible that X and Y are correlated because X causes Y or Y causes X or a third factor causes the correlation between X and Y. The gist of the nature-of-nurture phenomenon is that genetics is a third factor that causes correlations between 'environmental' measures and psychological traits.

In contrast, correlations between a polygenic score and a trait can only be interpreted causally in one direction – from the polygenic score to the trait. For example, we have shown that the educational attainment polygenic score correlates with children's reading ability. This correlation means that the inherited DNA differences captured by the polygenic score cause differences between children in their school achievement, in the sense that nothing in our brains, behaviour or environment can change inherited differences in DNA sequence.

In this way, polygenic score correlations eliminate the usual uncertainty about what is cause and what is effect when two variables are correlated. However, the correlation between a polygenic score and a psychological trait does not tell us about the brain, behavioural or environmental pathways by which the polygenic score affects the trait. A long slog up these pathways will be required to understand the intervening processes, especially because tens of thousands of DNA differences are involved, each with very small and highly pleiotropic effects. It is remarkable that polygenic scores can predict psychological traits without knowing anything about these intervening processes.

The second unique benefit of polygenic scores is that they can predict just as well from birth as they can later in life. Because inherited DNA differences do not change from cradle to grave, a person's polygenic score does not alter throughout the course of their life. In other words, if we had DNA from ourselves as infants and again as adults, the SNP genotypes would be identical and so too would the infant and adult polygenic scores. For this reason, polygenic scores can predict adult traits from infancy just as well as from adulthood.

In contrast, there is nothing else that can tell us if an infant is going to get a PhD or a psychosis. Infants' psychological characteristics, such as their temperament and cognitive development, tell us little about what infants will be like as adults. Even for intelligence, the most predictive psychological trait, no traits in the newborn predict later intelligence. When infants are two years old, intelligence tests predict less than 5 per cent of the variance of scores when the individuals are eighteen years old. In contrast, polygenic scores can predict just as much variance in adult intelligence as they can, not just at two years of age, but even at birth.

The third unique feature of polygenic scores is that they can predict differences between family members. Before the DNA revolution, genetic predictions were limited to estimates of family resemblance. For example, your risk of schizophrenia is 9 per cent if you have a first-degree relative who has been diagnosed as schizophrenic, a ninefold increased risk, as compared to the population risk of 1 per cent. This prediction is the same for all children in a family. But genetic risk is not the same for all children in a family because siblings are 50 per cent different genetically (unless they are identical twins).

Polygenic score predictions are specific to an individual, not general to a family. This means that a polygenic score for schizophrenia can show that one sibling has a greater vulnerability than another sibling. Or that one sibling has a higher polygenic score for educational attainment, which could help parents understand why that sibling finds school much easier. Polygenic scores will expose the wide range of genetic differences between siblings. Polygenic score differences are just as great between parents and their offspring as they are for siblings. Children are only 50 per cent chips off the old block.

These unique features of polygenic scores will transform clinical psychology by changing the way we identify, treat and think about psychological problems. Specifically, polygenic scores will make a difference in five ways.

For the first time in psychology, polygenic scores will make it possible to identify problems on the basis of causes rather than symptoms. In psychology, problems are identified solely on the basis of symptoms, after the problem begins to make itself known. For example, depression is diagnosed by asking people about symptoms of depression, such as sadness, hopelessness and lack of enjoyment. Learning disabilities are diagnosed by poor performance on cognitive tests.

Not a single psychological problem is identified on the basis of causes rather than symptoms. People can of course be depressed for many reasons, but polygenic scores can predict the extent to which individuals are depressed for genetic reasons.

A second way in which polygenic scores will transform clinical psychology is by moving away from diagnoses and towards dimensions. One of the big findings in this book is that the abnormal is normal, meaning that, from a genetic perspective, there are no qualitative disorders, only quantitative dimensions. This finding comes from research showing that genetic risk for psychological problems is continuous from low to high risk. There is no point at which genetic risk tips over into pathology. It's all quantitative – a matter of more or less.

Polygenic scores provide unambiguous proof that genetic influence is continuous. Because polygenic scores aggregate thousands of DNA differences, they are perfectly normally distributed as a bell-shaped curve. Even when GWA studies are based on differences between diagnosed cases versus controls, polygenic scores derived from these case-control GWA studies are also normally distributed. This means that they not only predict whether someone is at risk for the disorder or not, they also predict variation throughout the distribution – from people who are frequently or severely depressed to people who are seldom depressed. Individuals whose polygenic scores are at the 20th percentile will be less depressed on average than those at the 40th percentile, who, in turn, will be less depressed on average than those at the 60th percentile.

The abnormal is normal, in the sense that we all have many of the thousands of DNA differences that contribute to the heritability of any psychological problem. Our risk depends on how many of these

DNA differences we have. Polygenic scores will contribute to the demise of diagnoses because polygenic scores make it clear that genetic risk is continuous, not dichotomous. It is worth repeating once again: There are no disorders to diagnose and there are no disorders to cure. Polygenic scores will be used to index problems quantitatively rather than deciding whether someone 'has' a disorder.

A third transformative impact of polygenic scores is to move clinical psychology from one-size-fits-all treatments to individually tailored treatments. Polygenic scores will really take off in clinical psychology as soon as we discover treatments that interact with genotypes, in the sense that the success of treatments depends on polygenic scores. Treatments can then be tailored to individuals on the basis of their polygenic scores. For example, profiles of polygenic scores might be used to predict whether a depressed person will respond better to talking therapies or drugs, or to a certain type of talking therapy or drug.

Individually tailored treatments have received the most attention in medical research using an individual's genotype to select appropriate drugs, known as *pharmacogenomics*. More generally, 'precision medicine' or 'personalized medicine' is a model for customizing healthcare on the basis of genetic or other biological information. The goal is to identify the most effective treatments for an individual, sparing the expense, side effects and wasted time for those who will not benefit from the treatment.

The fourth way in which polygenic scores will change clinical psychology is by shifting the focus from treatment towards prevention. As Benjamin Franklin said, an ounce of prevention is worth a pound of cure. In psychology as well as medicine, we have had to wait for problems to occur and then try to fix them. Many psychological problems, such as alcohol dependence and eating disorders, are difficult to cure once they become full-blown problems, in part because they cause collateral damage that is difficult to repair. Preventing problems before they occur is much more cost effective economically, as well as psychologically and socially.

Prediction is the *sine qua non* for prevention and polygenic scores are the perfect early-warning system. They can predict from birth just as well as later in life. In addition, polygenic scores are not just biomarkers – their prediction is causal.

Although we know surprisingly little about specific interventions to prevent the emergence of psychological problems, polygenic scores will facilitate research on prevention because, for the first time, polygenic scores make it possible to identify individuals who are at risk. For example, for depression, some treatments seem likely to be useful as preventive interventions. Cognitive behavioural therapy and well-being training seem obvious candidates to prevent depression as well as alleviate its symptoms. However, the effects of large-scale preventive programmes administered in schools, in the community or on the internet are small and temporary. We cannot afford intensive and expensive preventive interventions for everyone, but if we can target individuals at high genetic risk it would be cost effective to intervene at a personal level, for example, providing extended one-on-one cognitive behavioural therapy. Polygenic scores make the possibility of targeted prevention a reality.

Another example is attention deficit hyperactivity disorder. There have been attempts to prevent hyperactivity by giving parents instruction and creating education programmes and preschool programmes based on playing games, but success so far has been modest. Again, it seems likely that you get what you pay for. More intensive, and thus more expensive, interventions have a better chance of success, but this would only be feasible if we can identify children at high risk. And now we can do this, using polygenic scores.

The fifth transformative feature of polygenic scores is that they will promote *positive genomics*. As we have seen, polygenic scores are always perfectly normally distributed, which means that both ends of the distribution are the same size. Clinical psychology focuses on the negative end of the distribution – the problems, disabilities and vulnerabilities. Polygenic scores, on the other hand, will inspire a switch of focus to the other, positive, end of the distribution – strengths instead of problems, abilities instead of disabilities, and resiliencies instead of vulnerabilities.

The positive end of the polygenic score distribution should not be defined as merely low risk. It is possible that this 'other end' of the distribution of polygenic scores for psychopathology has its own problems. The word 'risk' should be avoided in relation to polygenic scores because it misses this deeper meaning of polygenic scores implied by their normal distribution. For example, my low polygenic score for bipolar disorder might mean something other than being at low risk for the disorder. It might mean that I am flat in affect, failing to experience the highs and lows of life. Using hyperactivity as another example, a high polygenic score will predict impulsiveness and inattentiveness, although no polygenic score is currently available. Does a low score just imply a low risk for being impulsive and inattentive? Or does it predict the opposite problems of being compulsive and obsessive? Similarly, the low end of the polygenic score for BMI might not just predict low risk for obesity. It might signal fussiness about food that leads to eating disorders like anorexia.

As these examples suggest, it is possible that, when it comes to polygenic scores for disorders, intermediate scores are better than extremely low scores. Everything in moderation, as my mother used to remind me, without effect. (Mothers matter, but they don't make a difference.) I always preferred Oscar Wilde's take: 'Everything in moderation, including moderation.'

Because polygenic scores are so new, next to nothing is known about the 'other end' of the normal distributions of polygenic scores for disorders. In addition to stimulating research on positive genomics, polygenic scores will foster the promotion of health, in addition to the prevention of illness. For cognitive traits, polygenic scores will shift research from disabilities to abilities, including promotion of high ability.

Clinical psychology will be changed beyond recognition by polygenic scores, which focus on causes instead of symptoms, dimensions instead of diagnoses, individually tailored rather than one-size-fits-all treatments, prevention instead of treatment, and a positive emphasis on health rather than illness.

Polygenic scores will also revolutionize psychological research. For forty years I have been trying to understand what causes people to differ so much in their psychology, beginning with the fundamental question of the relative importance of nature and nurture. Research has consistently shown that inherited genetic differences account for the bulk of psychological differences, especially systematic differences, between individuals.

For the last twenty years I hoped to move from the indirect genetic methods of twin and adoption studies to methods that assess inherited DNA differences directly for individuals. That has finally happened, and it feels like winning a twenty-year-rollover lottery. It is exciting to see how quickly the DNA revolution is transforming research in psychology. Polygenic scores make it possible for researchers to ask questions that go beyond nature versus nurture with far greater precision and sophistication. They will also democratize genetic research in psychology by making it possible for any researcher to incorporate genetics into their research on any topic with any sample, as long as they collect DNA. No longer is the price of admission special samples like twins and adoptees.

One set of questions is about development. A polygenic score derived from a GWA study of adults – schizophrenia or educational attainment, for example – can predict adult schizophrenia or educational attainment from birth just as well as in adulthood. But how early in development can a polygenic score predict differences in children's behaviour? Studies of children at genetic risk because one of their parents was diagnosed as schizophrenic have not been able to find any physiological or psychological markers of schizophrenia before adolescence. However, polygenic scores will provide greater resolution than family risk for finding problems early in development that might be targets for intervention and prevention.

For educational attainment, we have seen that the polygenic score created from a GWA study of years of education in adults can predict 9 per cent of the variance in tests of school achievement at the age of sixteen. How early can this polygenic score predict children's school achievement? We found in TEDS that the educational attainment polygenic score predicts 5 per cent of the variance in school achievement in secondary school at the age of twelve, and it even predicts 3 per cent of the variance in primary school at the age of seven.

I find it incredible that a polygenic score derived from a GWA study that analysed the coarse variable of number of years of education for adults is able to predict children's achievement even in the early school years. These results imply that a GWA study focused on children's achievement at school could produce polygenic scores that predict several times more variance, although no such GWA studies have yet been reported.

A second set of questions follows from the big finding of generalist genes. That is, instead of distinct sets of genes for schizophrenia and bipolar depression, twin studies suggest that many of the same genes affect both. The same generalist-genes phenomenon has been found for apparently different cognitive abilities like verbal ability and memory. Polygenic scores will foster multivariate research because, once SNP genotypes are available, it is easy to create dozens of polygenic scores.

GWA studies have found genetic correlations greater than 0.5 between schizophrenia, major depressive disorder and bipolar disorder in the PGC, which we replicated in TEDS. An exciting new challenge for research is to understand what this general genetic factor of psychopathology is, how it develops and its implications for treatment and prevention.

The educational attainment polygenic score has already shown its general effects across diverse psychological traits. As we have seen, it predicts 4 per cent of the variance in the target trait of years of education in adults, but it predicts even more variance in other traits, such as tested school achievement (9 per cent), intelligence (5 per cent), and comprehension and efficiency of reading (5 per cent). The power of the educational attainment polygenic score comes from its large GWA sample size. Its ability to predict intelligence and reading comes from generalist genes. The combination of these two factors is why it predicts more variance in intelligence than GWA studies that specifically targeted intelligence.

Although it has been surprising to see how general genetic effects are on mental illness and mental abilities, there are of course trait-specific genetic effects, for example SNPs specific to schizophrenia or reading. An important direction for research is to create trait-specific polygenic scores as a counterpoint to research on generalist genes. Trait-specific polygenic scores might be more amenable to trait-specific intervention and prevention.

A third set of questions is about the interplay between nature and nurture. The big finding from twin studies can be summed up as the nature of nurture, which refers to discovering genetic influence on environmental measures such as life events, parenting and peers. Because genetics influences environmental measures as well as psychological measures, genetics is also responsible in part for correlations between environmental measures and psychological measures.

Polygenic scores can be used to nail down genetic influence on the variance of environmental measures and on their covariance with psychological measures. They can also control for genetic influence in order to study purer environmental effects. For example, in research correlating the family environment with children's cognitive development, such correlations can be corrected for the polygenic score for educational attainment as a partial control for genetic influence.

Polygenic scores also make it possible to study the interplay between nature and nurture between families rather than within families. That is, twin studies can only look at experiences that differ for children in a family, for example, whether their parents are more loving to one child than to another. This focus on differences within families misses how loving the parents are compared to other parents, that is, differences *between* families rather than *within* families. In other words, even if a parent is more loving towards one child than to another, the parent might not be very loving to either child, as compared to other parents.

Unlike twin analyses, a polygenic score for a child can be used to investigate the nature of nurture between families as well as within families. For example, one of the best 'environmental' predictors of children's school achievement is socioeconomic status, which is intrinsically a between-family measure. That is, children within a family obviously experience the same socioeconomic status. A twin study would not make sense here because twins in a family experience the same socioeconomic status. Identical and fraternal twin correlations would both be 1 because there are no differences within families, so heritability would be 0 and shared environmental influences would be 100 per cent.

Although socioeconomic status is often assumed to be a purely environmental measure, the nature-of-nurture finding suggests that we should expect genetic influence on any measure of the environment. Moreover, the major component of socioeconomic status of parents is their years of education. So, it should come as no surprise that we have found that the educational attainment polygenic score correlates with parents' socioeconomic status.

Another twist is that children's own educational attainment polygenic score correlates almost as much with their parents' socioeconomic status. What's more, it also accounts for half of the correlation between family socioeconomic status and children's school achievement, meaning that the correlation is mediated genetically. These results are surprising only if you think that socioeconomic status is a purely environmental variable.

The educational attainment polygenic score also mediates correlations between other 'environmental' predictors and school achievement. For example, breastfeeding correlates positively with children's school achievement and watching television correlates negatively. We have shown that the polygenic score for educational attainment explains a significant portion of the correlation between both of these 'environmental' measures and children's school achievement, meaning again that this correlation is in part mediated genetically.

These are all DNA examples of the nature of nurture, the first studies of this type using polygenic scores. The evidence from twin studies suggests that genetics accounts for about a third of the variance of environmental measures. This phenomenon is called genotype–environment correlation because it literally means that there is a correlation between genotype – in this case, a specific polygenic score – and environment. Genotype–environment correlation suggests a new way of thinking about experience, that is, how genes use the environment to get what they want. Genotype–environment correlation provides a general model for how genotypes become phenotypes; that is, how we select, modify and create environments correlated with our genetic propensities.

Another type of interplay between genes and environment sounds similar but is actually very different. Genotype–environment interaction is not about the correlation between genes and environments but their interaction. That is, does the effect of the environment depend on an individual's genotype? For example, does the effect of being bullied depend on a child's genotype? Genotype–environment interaction is about different strokes for different folks. It is the essence of precision psychology, which aims to tailor treatments to individuals, not relying on one-size-fits-all approaches. In education, this is at the heart of personalized learning.

Eagerness to find genotype–environment interactions led to early attempts to identify interactions between candidate genes and environments as they affect psychological traits. The earliest and most famous report of genotype–environment interaction involved an interaction in which a candidate gene's association with antisocial behaviour showed up only for individuals who had suffered severe childhood maltreatment. Many other interactions between candidate genes and psychological traits have been reported, but most have not replicated. Polygenic scores will re-energize the search for genotype–environment interaction.

Although research on genotype–environment interaction using polygenic scores can study the interplay between any environmental measure and any psychological trait, a focus for this research will be individually tailored treatments for psychological disorders. We do not yet have a polygenic score that predicts differential responses to psychological treatment but, if a powerful polygenic score were developed, it would be in demand.

Polygenic scores will be valuable for looking at these traditional questions about development, links between traits, and gene–environment interplay. But the most exciting aspect of polygenic scores is the potential they offer for completely new and unexpected directions for research. I will mention three examples from my team's current research. None of this work could have been done without the educational attainment polygenic score.

The first example seems shocking: Children in private and grammar schools in the UK have substantially higher educational attainment polygenic scores than students in comprehensive schools. In the UK, private schools are privately funded and grammar schools are state-funded but what they have in common is that they both select their students. Comprehensive schools are state-funded but are not allowed to select students.

How is it possible that students in private and grammar schools differ in their DNA from students in comprehensive schools? The answer is not surprising if you recall the results of the TEDS study that showed that students in selective secondary schools get better GCSE scores on average than students in non-selective secondary schools simply because selective schools select students more likely to achieve better scores in the first place, not because of value added by the selective schools. Selective schools select students on the basis of previous school achievement in primary school and standardized tests of intelligence, so it is a self-fulfilling prophecy that these students do better in secondary school.

After controlling for these selection factors, there is no difference in achievement. The factors on which students are selected – primarily prior achievement and intelligence – are substantially heritable. Therefore, it is not surprising that the GCSE difference between selective and non-selective schools is heritable, and this is what is reflected in our finding that the average educational attainment polygenic score is higher in students in selective as compared to non-selective schools.

This is another example of one of the big findings from genetic research, the nature of nurture. Private versus public schooling is assumed to be an environmental factor, but the differences in school achievement are actually genetic in origin. That is, children apply to and are accepted by selective schools for genetic reasons.

An implication for parents is that it is not worth the huge amount of money needed to send children to private school if you are doing it because you think it will improve their school achievement. Even if you accept that private schools do not make a difference academically, you might think that private schooling improves children's chances in other ways, such as going to a better university, making better career choices and earning a higher salary. These outcome differences exist, but they are also largely due to pre-existing student characteristics, meaning that these students would have done as well if they had not gone to private schools. Although these conclusions may not be easy to swallow, they follow from this book's general finding that inherited DNA differences are the major systematic force making us who we are.

The second example of new research directions involves what is called intergenerational educational mobility, specifically whether children have equal opportunities to go on to higher education, regardless of whether their parents did. The best predictor of whether children go to university is whether their parents went to university, a link which is widely assumed to be environmental in origin and which is thus thought to be a sign of immobility and lack of equality. In other words, university-educated parents are thought to pass on environmental privilege to their children, creating inequality in educational opportunity and stifling intergenerational educational mobility. In comparisons between countries, the strength of this link between parent and offspring attainment is used as an index of educational inequality and the lack of social mobility.

However, what we are talking about here is parent–offspring resemblance for educational attainment. I hope that by now you find it odd that people have assumed that parent–offspring resemblance is caused environmentally and that possible genetic influence has not been considered. Using the TEDS dataset, we found that DNA differences underlie this parent–offspring resemblance. That is, educational attainment polygenic scores of children were highest when both parents and their children went to university and lowest when neither parents nor their children went to university. Finding genetic influence on parent–offspring resemblance for educational attainment is not surprising. A substantial body of research has shown that educational attainment is heritable. Indeed, years of education was the target trait for the GWA study that resulted in the educational attainment polygenic score.

The novel aspect of these findings is that genetics drives *differences*, not just similarities, in educational outcomes between parents and their children, which is a key index of mobility. We looked at the polygenic scores of upwardly mobile children; that is, those who went to university even though their parents did not. We found that these upwardly mobile children have higher educational attainment scores than children who, like their parents, did not go to university. In other words, genetics gives some children born into socially disadvantaged families a chance to overcome the constraints of their background, as long as there is mobility. Regardless of where parents' scores lie in the distribution, their children will have a wide range of educational attainment scores. Social mobility means that children with the genetic propensity to do well at school will have the opportunity to perform to the best of their ability, regardless of their environmental background.

Downward mobility is also governed by genetics. Children whose parents went to university are less likely to go to university if the children have lower educational attainment polygenic scores. Finding genetic influence on downward mobility as well as upward mobility is important because it is the first step towards preventing the creation of genetic castes.

Our twin analyses backed up these polygenic-score findings by showing genetic influence on both upward and downward mobility. Identical twins were more likely than fraternal twins to be similar in their upward or downward mobility. These analyses suggested that genetics accounts for about half of the individual differences in upward and downward mobility.

Overall, these findings turn current thinking about social mobility and educational opportunity on their head. Parent–offspring resemblance for educational attainment primarily reflects genetic influence, not environmental inequality. This is another example of the conclusion that heritability, in this case parent–offspring resemblance, is an index of equality of opportunity, as discussed in Chapter 9. Greater reduction in environmental inequalities of privilege, wealth and discrimination will result in greater heritability of educational outcomes.

Upward mobility is likely to be a pleasant surprise for parents who were not university educated and who see their child blossom intellectually. This was definitely the case for my parents, who did not go to university and were pleased and proud that I did. Conversely, downward mobility is difficult for university-educated parents to accept. Polygenic scores might help these parents recognize that a child's lack of interest in higher education is not necessarily a sign of recalcitrance or laziness. Instead, the child might not have the aptitude or appetite for higher education for genetic reasons.

It is worth repeating that genetics should foster a recognition and respect for individual differences. Genetic influence does not imply hard-wired programming that you can't change. But, when possible, it makes sense to go with the grain of genetics rather than against it. Using university education as an example, parents could pull out all the stops to get a child into university against their genetic propensities, but this could come at a cost if higher education doesn't suit them.

The third and final example of new research directions involves changes in heritability following major societal change. As a reminder, heritability describes the relative influence of DNA differences and environmental differences in a particular population at a particular time. Like all descriptive statistics, such as means, variances and correlations, heritability will change as the population changes.

One type of change was implied in the earlier discussion of meritocracy. Heritability can be viewed as an index of success in achieving meritocratic values of equality of opportunity by rewarding talent and effort, rather than rewarding environmentally driven privilege. Talent and effort are substantially influenced by genetic factors. This suggests that socioeconomic status should be more heritable as a country becomes more meritocratic. As environmentally driven differences decline, genetic differences account for more of the remaining differences in socioeconomic status.

Estonia provided an opportunity to test the hypothesis that the heritability of educational attainment and occupational status increases with greater meritocracy. In 1991, as the Soviet Union dissolved, Estonia became independent and quickly moved away from the centralized and politicized reward system of the Soviet Union towards more meritocratic selection of individuals for education and occupation. If greater meritocracy leads to greater heritability of socioeconomic status, we would predict that the educational attainment polygenic score relates more strongly to socioeconomic status after independence.

As often happens in research, testing this hypothesis was made possible by fortuitous events. First, Estonia has been at the leading edge of the DNA revolution, as well as other technological advances. The Estonian Genome Centre at the University of Tartu created a databank that includes DNA, SNP chip genotypes and extensive data on more than 50,000 Estonians, which is 5 per cent of the adult population, and they are now adding another 100,000 participants. A second fortuitous factor was that one of my graduate students was from Estonia and she facilitated a collaboration that allowed us to test the hypothesis.

We found impressive confirmation of the hypothesis. The educational attainment polygenic score predicted twice as much variance of educational attainment and occupational status in the post-Soviet era. Increased genetic influence for occupational status was especially great for women, which makes sense, because women had the most to gain from meritocracy.

This finding is another example of how heritability can be seen as an index of equality of opportunity and meritocracy.

Polygenic scores have made an impressive debut in psychology, already becoming our best predictors of schizophrenia and school achievement. There is a long way to go until they reach their full potential of predicting all of the 50 per cent heritable variance in psychological traits. Given how fast-paced research is in this field, it seems safe to predict that we will eventually have polygenic scores that predict hefty chunks of variance for all psychological traits – mental health and illness, mental abilities and disabilities, personality and the scores of other traits, like attitudes and interests. Polygenic scores will be the best predictors of these traits because inherited DNA differences are the main systematic force in making us who we are.

Despite their novelty, polygenic scores are already transforming clinical psychology and psychological research in general. In closing, I would like to speculate about how polygenic scores will affect all of us as we enter the era of personal genomics, looking forward a few years to a time when we have many more, and much more powerful, polygenic scores for psychological traits. I should acknowledge in advance that some of these speculations will be highly controversial. I am speculating about what I think might happen and why. I am not advocating that these things happen but raising them as issues that need to be discussed.

The public source of polygenic scores will be direct-to-consumer companies which will soon add polygenic score profiles to the single-gene genotyping and ancestry data which they currently provide to millions of people. My psychological polygenic scores provide a glimpse of how this information can be useful for self-understanding, as well as a look at the limitations of prediction at an individual level.

Self-understanding is relatively benign, although even this raises some concerns, considered earlier.

However, self-understanding only scratches the surface of applications of psychological polygenic scores. Other applications are more vexing, psychologically as well as ethically. For example, it seems likely that parents will soon be able to obtain polygenic scores for their children, perhaps at birth, to tell their children's genetic fortunes. I think many parents will be motivated to do this simply out of curiosity, an extension of self-understanding, despite concerns that have been raised about violations of children's privacy and possible self-fulfilling prophecies caused by labelling. Although parental curiosity about their children's future might seem frivolous or even dangerous, good can come from parents getting a genetic glimpse of their children's individuality – their strengths and weaknesses, their personalities and their interests. This information might help parents to try to maximize their children's strengths and minimize their weaknesses.

Anne Wojcicki, not an unbiased commentator because she is the founder of 23andMe, argues that it is the duty of parents to arm themselves with their child's genetic blueprint, and her company makes it as easy for parents to obtain genomic information for their children as for themselves. There are many examples of how polygenic-score information could be useful to intervene to prevent problems, or at least to be forewarned about them. For example, polygenic scores will be able to predict reading disability. Rather than waiting until a child gets to school and fails to learn to read, being able to predict that the child is likely to have a problem learning to read gives parents the opportunity to intervene earlier to ward off the problem. At the least, a high polygenic score for reading problems will alert parents that their child might need extra help in learning to read. Moreover, most children who find it difficult to learn to read have earlier problems with language, so parents might intervene to stimulate language learning before children begin to read.

Many other examples come to mind about using polygenic scores to make life easier for children. For children whose polygenic scores suggest that they are prone to depression, we can help them use the strategies of cognitive behavioural therapy, such as avoiding rumination about problems and approaching difficulties in a more positive way by breaking them down into smaller parts. For personality, there are common-sense things a parent can do. Knowing that a child has a high energy level can help parents realize that their child needs opportunities to burn off some of that energy. A shy child could be helped by being eased into situations with strangers.

The most alarming prospect for many people will be the potential use of polygenic scores by parents to choose an embryo with the 'best' polygenic profile score. There have long been concerns about the possibility of 'designer babies'. The need for this decision can emerge when several viable embryos are created during the process of in vitro fertilization, one of many types of assisted reproduction. It seems unlikely that a couple would go through the unpleasant process of in vitro fertilization solely for the purpose of selecting an embryo on the basis of its psychological polygenic score profile. More likely, a couple would undergo in vitro fertilization for medical reasons, for example, to screen for single-gene recessive disorders when the couple are both carriers, or because they have struggled to conceive. A classic ethical conundrum is to ask what you would do if you had several equally viable embryos but you could only implant one. If we had to make such a choice, it would seem obvious to avoid selecting an embryo with a major single-gene medical disorder. But if there were a further choice to make, would you look at physical, physiological and psychological polygenic profiles?

Polygenic score profiles could have an impact even earlier in the life cycle, before reproduction, in selecting a mate. Genetic selection is already happening at a single-gene level to make it possible for couples to find out if they are both carriers for any of the thousands of single-gene recessive disorders. If they are both carriers, this would mean that their children would have a 25 per cent chance of having the disorder. Carrier screening is worth considering for prospective couples because, although these single-gene disorders are rare, carriers are common. For example, phenylketonuria (PKU), a single-gene recessive disorder which, left untreated, causes severe intellectual disability, occurs in only one in ten thousand people, but one in fifty of us is a carrier. Thus, there is a significant chance that a couple are both carriers for one of the thousands of single-gene disorders. They could decide not to have children to avoid this risk or be forewarned of problems they will face if they had an affected child. They could also consider other options, such as undergoing in vitro fertilization to screen for that one-in-four possibility.

Although it might seem far-fetched and perhaps dystopian, dating websites might extend their data to include polygenic scores. As research on polygenic scores progresses, it could become possible to include polygenic scores for psychological traits typically included on dating websites such as mental health, intelligence, earning potential, ambitiousness, physical fitness, personality traits and relationship qualities – and even good sense of humour. Unlike the hype of dating websites, polygenic-score information could be verifiable through password-protected links to a direct-to-consumer company that provides specified polygenic scores. Whether greater control over selection of a mate increases the long-term prospects of a couple remains to be seen, however.

These potential applications involve our personal use of our own genomic data. What about the use of our genomic data by others? In medicine, this is acceptable – in fact, it is the goal of precision medicine. But what if psychological polygenic scores became part of the selection process for education and employment? This is the nightmare scenario for many people; the 1997 film *Gattaca* reverberates in the public consciousness as a dystopian vision of a world divided by DNA into the 'valids' with ideal genomes who are in charge, and the 'in-valids' serving as a genetic underclass. *Gattaca*'s view of a world dichotomously divided by DNA into valids and in-valids misses the point that polygenic scores are always perfectly normally distributed – they are dimensional, not dichotomous. Most of us are in the middle.

However, *Gattaca* touched a nerve because it warned about the dangers of genetic information in the hands of a totalitarian state. But there is another way to look at it in democratic societies, especially ones that favour meritocracy. We already administer psychological tests in order to select people for education and, to a lesser extent, for employment. If we are going to select people, the predictive power of polygenic scores could supplement information we already obtain from testing. In addition to their predictive power, polygenic scores are more objective and free of biases like faking and training, as compared to tests that we currently use for selection. You can't fake or train your DNA.

The usefulness of polygenic scores in the context of selection is an empirical issue, although utility does not address ethical concerns, which I will consider later. We have seen, even in these early days of research, that polygenic scores can usefully supplement test scores to predict achievement at secondary school and university. Polygenic scores could be especially useful in pinpointing children whose disadvantaged backgrounds might otherwise reduce their opportunities for higher education. Another example of the potential good that can come from polygenic scores is to consider underachievement and overachievement in terms of discrepancies between potential performance predicted by polygenic scores and actual performance. More generally, polygenic scores are key for personalized learning, as they predict pupils' profiles of strengths and weaknesses, which offers the possibility to intervene early to prevent problems and promote promise.

For selection for the purpose of employment, it is again an empirical issue how much polygenic scores can add to the prediction of success on the job. It seems likely that polygenic scores can help because tests and interviews are notoriously poor at predicting job success, predicting just a few per cent of the variance. Psychological polygenic profiles might be especially useful in considering patterns of strengths and weaknesses that predict success at particular jobs. Similar to the example of dating websites, a password-protected link to a direct-to-consumer company could make available a certified set of polygenic scores relevant to occupational selection in general and different sets of polygenic scores relevant to specific jobs.

As scary as some of these possibilities might seem, I predict that they will happen eventually. Given *Gattaca*-type concerns, let's consider the bête noire of genetic screening of newborns in greater detail. Even though newborns cannot provide informed consent, we have nonetheless genetically screened newborns for decades – it is compulsory in most countries. The original reason for screening newborns was phenylketonuria (PKU), a single-gene disorder that causes severe intellectual disability in about one in ten thousand babies and accounts for 1 per cent of the intellectually impaired population in institutions.

PKU involves a mutation in a gene that breaks down phenylalanine, an essential amino-acid building block of proteins. Our bodies do not produce phenylalanine; we get it from many protein-laden foods, at first from breast milk and later from meat and cheese. To use phenylalanine, we need to metabolize it. Individuals with PKU have a malfunctioning enzyme which causes unprocessed phenylalanine to build up, and this damages the developing brain. Untreated, the PKU mutation causes severe cognitive impairment. Over 80 per cent of those with untreated PKU require twenty-four-hour support and 70 per cent cannot talk beyond single words.

For forty years, newborns around the world have had their heel pricked to get a drop of blood to test for PKU. This quick and inexpensive test, called the Guthrie test, assays the gene's protein product for tell-tale signs of PKU. The reason why newborns have been screened for this rare genetic disorder is that the worst effects of PKU can be prevented with a low-tech, inexpensive intervention. But this bullet can be dodged only if the intervention begins early in life. Because children with PKU cannot metabolize phenylalanine, which thus builds up and damages the developing brain, it makes sense that a simple solution is to limit the intake of phenylalanine with a diet low in phenylalanine.

The decision to screen depends on the ratio of benefits to cost. The benefit-to-cost ratio for PKU is so huge it seems unimaginable not to screen. The cost of screening is negligible compared to the psychological cost to parents and the economic cost to society of life-long care. In stark contrast, the low-tech, low-cost dietary intervention for PKU changes this bleak prognosis into one of a nearly normal life.

There are as yet no other genetic stories like PKU with such happy endings. Nonetheless, newborns are now screened at the same time for dozens of other single-gene disorders, including cystic fibrosis and congenital hypothyroidism. The point here is that we have been screening newborns for genetic disorders for a long time. So it's not a question of whether we do it but rather how much we do it. Why screen only for a few genetic mutations, instead of thousands of known single-gene disorders? Why not obtain polygenic scores to predict common problems, including psychological problems? Using SNP chips or, even better, whole-genome sequencing, the cost would be about the same as screening separately for a few genetic mutations.

The use and abuse of psychological polygenic scores also comes down to cost–benefit analyses, where the costs and benefits are as much psychological as they are medical and economic. The complexity of these analyses is that they yield different results depending on whether the perspective taken is that of the child, the parent or society. Moreover, there are individual differences in these cost–benefit analyses because people differ in their perceptions of the costs and benefits about knowing versus not knowing their genetic future. The costs of personal genomics have been widely discussed, particularly in relation to single-gene medical disorders. These include concerns about privacy, discrimination, stigmatization and designer babies. Another issue is the emotional impact of genomic knowledge, not just to the person who signed up to get their polygenic scores but also to relatives for whom the information is also relevant but who did not sign up.

A wise move from the beginning of the Human Genome Project, which sequenced the human genome, was to use a part of the budget to fund research on the ethical, legal and social implications (ELSI) of the project. The ELSI programme has addressed many of these issues at the level of single-gene causes of medical disorders, such as privacy and fairness in the use of genetic information, as well as the integration of genetic testing into clinical settings, ethical issues surrounding the design and conduct of genetic research, and the professional and public understanding of the complex issues that result from genomic research.

I hope that these vexing issues of the costs of personal genomics will be worked out at this level of single-gene medical disorders. These issues are not as severe when it comes to polygenic scores for common psychological disorders and dimensions because polygenic scores are inherently probabilistic rather than deterministic.

My general point is that polygenic scores represent a major scientific advance and, like all scientific advances, they can be used for good as well as for bad. I have highlighted their potential for good in psychology and society as an antidote to the dystopian doom and gloom that often permeate discussions of personal genomics. We need to discuss the pros as well as the cons so that we can maximize the benefits and minimize the costs, because the DNA revolution is unstoppable. Although there are many psychological and ethical issues to consider, millions of people have already voted with their credit card by paying to have their genomic fortunes foretold, even before polygenic scores are available. Genomics is here to stay. The internet has democratized information to such an extent that people will not tolerate paternalistic regulations that prevent them from learning about their own genomes. The genome genie is out of the bottle and, even if we tried, we cannot stuff it back in.

Epilogue

More than sixty years after the discovery of the structure of DNA and fifteen years after the human genome was sequenced, DNA has come to psychology. In this book we have traced the journey from genetics to genomics in psychology. The first stop along the way was to realize that DNA is the most important factor in making us who we are. Inherited DNA differences are the essence of human individuality. Over the past century research based on twin and adoption studies built a mountain of evidence documenting the importance of inherited DNA differences, which account for half of the differences between us, not just in our bodies but in our minds as well – for mental health and illness, personality and cognitive abilities and disabilities. Accounting for half of the variance in these complex traits is off the scale compared to any other effect sizes in psychology, which rarely account for 5 per cent of the variance, let alone 50 per cent.

Genetic researchers then went beyond demonstrating heritability to ask more interesting questions. How does genetic influence unfold during development? Are there genetic links between normal and abnormal development? Do different genes affect different dimensions and disorders? Two of the most fascinating questions were about nurture rather than nature. Genetically sensitive designs like twin and adoption studies could, for the first time, study the environment while controlling for genetics.

This research led to five of the biggest findings in psychology. Studying environmental measures in genetically sensitive designs led to the first discovery: Most measures of the environment used in psychology show substantial genetic influence. What look like environmental effects in correlations between 'environmental' measures and psychological traits are actually genetic effects.

The second is about development. Heritability increases throughout the lifespan, especially for intelligence. The third finding is about the substantial genetic links between normal and abnormal behaviour. The genetic links are so strong that the bumper-sticker summary of this research is that 'the abnormal is normal'. The fourth finding is about the robust genetic links between supposedly different traits, suggesting that genetic effects are general across traits rather than specific to each trait. Fifth, studying the environment while controlling for genetics revealed that environmental influences make children growing up in the same family as different as children reared in different families.

These findings led to a new view of what makes us who we are. Genetics accounts for most of the systematic differences between us – DNA is the blueprint that makes us who we are. Environmental effects are important too, but they are unsystematic and unstable, so there's not much we can do about them. Moreover, what look like systematic environmental effects are often due to us choosing environments correlated with our genetic propensities. Together, these findings suggest that parenting, education and life experiences don't make a difference in psychological traits, even though they matter tremendously. These findings also imply a new way of thinking about equal opportunity and meritocracy, in which higher heritability of educational attainment, occupational status and income is an index of greater equality of opportunity and meritocracy.

Just as the pace of discoveries like these was beginning to slow, along came the DNA revolution. Identifying all 3 billion bases in the double helix of DNA in the human genome uncovered millions of inherited DNA differences. Hundreds of thousands of single-nucleotide polymorphisms (SNPs) could be genotyped for an individual quickly and cheaply on a SNP chip.

Constructing a SNP chip with SNPs selected across the genome enabled genome-wide association (GWA) studies. GWA studies have been a game-changer for biological and medical sciences, as well as psychology. After a few faltering steps and stumbles GWA researchers made their first huge discovery about inherited DNA differences.

For complex dimensions and common disorders, including all psychological traits studied so far, the biggest effects of SNPs are incredibly tiny. This is why it was so hard at first for GWA studies to find associations between SNPs and complex traits. Effects as tiny as these can only be seen when GWA studies reach sample sizes of tens of thousands of cases for disorders such as schizophrenia, or hundreds of thousands of unselected individuals for dimensions like educational outcomes. As GWA studies reached these daunting demands for statistical power, they struck gold.

But what GWA studies found was gold dust, not nuggets. Each speck of gold was not worth much, but scooping up handfuls of gold dust made it possible to predict genetic propensities of individuals. Some plain dirt was scooped up too, but this doesn't matter, as long as we keep getting more gold. These polygenic scores mark the beginning of personal genomics in psychology in which our genetic futures can be foretold.

The first wave of polygenic scores, consisting of tens of thousands of SNP associations from GWA studies, can predict 17 per cent of the variance in height, 6 per cent of the variance in weight, 11 per cent of the variance in school achievement, 7 per cent of the variance in intelligence, and 7 per cent of the variance in liability to schizophrenia. Polygenic scores are already the best predictors we have for schizophrenia and school achievement. Most importantly, unlike any other predictors, polygenic scores predict just as well from birth, and their prediction is causal, in the sense that nothing changes inherited DNA differences.

Wave after wave of polygenic-score research is coming in, and each wave brings us closer to the high-water mark that will identify all the DNA variants responsible for heritability. Right now, the tide falls far short of the high-water mark of heritability, in part because the specks of gold dust are so small they are difficult to find. Nonetheless, by the time you read this, the predictive power of all of these polygenic scores will be far greater than those described in this book. The only way is up.

Before polygenic scores appeared genetic research showed us that heritability is substantial and ubiquitous for psychological traits, but this was only a general statement that could not be translated to genetic predictions for an individual. Now, polygenic scores are transforming clinical psychology and psychological research because DNA differences across the genome can be used to predict psychological traits for each and every one of us.

No doubt some of these findings and their interpretation will be controversial. People worry about change, and polygenic scores will bring some of the biggest changes ever, as the DNA revolution sweeps over psychology in waves of polygenic scores. Although we touched on some of the concerns about the applications and implications of this new frontier, I am excited about these changes because they are full of potential for good, and we can avoid the hazards if we are alert to them.

Now is the time to launch a broader public conversation about the applications and implications of the DNA revolution in psychology, because it will affect all of us. The main reason I wrote this book was to foster this discussion and to provide the DNA literacy that we need to address these complex issues in an informed way. Genetics is much too important to leave to geneticists alone.

Notes

PART ONE: WHY DNA MATTERS

Chapter 1: Disentangling nature and nurture

p. 3 *'For most of the twentieth century it was assumed that psychological traits were caused by environmental factors . . . called nurture'*: Steven Pinker, *The Blank Slate: The Modern Denial of Human Nature* (Penguin, 2003).

p. 5 *'2017 survey of 5,000 young adults'*: Emily Smith-Woolley and Robert Plomin, *Perceptions of Heritability*. Manuscript in preparation.

p. 5 *'Even though innate characteristics are programmed by DNA, we can't talk about their heritability because innate characteristics do not vary between us'*: This point was made well by geneticist and evolutionary biologist Theodosius Dobzhansky, who was the first president of the Behavior Genetics Association: 'The nature–nurture problem is nevertheless far from meaningless. Asking right questions is, in science, often a large step toward obtaining right answers. The question about the roles of genotype and the environment in human development must be posed thus: To what extent are the differences observed among people conditioned by the differences of their genotypes and by the differences between the environments in which people were born, grew and were brought up?' Theodosius Dobzhansky, *Heredity and the Nature of Man* (Harcourt, Brace & World, 1964, p. 55).

p. 6 *'Table 2. How much are these traits influenced by genetics?'*: The main reference for these results is my behavioural genetics textbook, Valerie Knopik et al., *Behavioral Genetics, 7th edition* (Worth, 2017). References for some of the newer data follow. Remembering faces: Nicholas Shakeshaft and Robert Plomin, 'Genetic Specificity of Face Perception', *Proceedings of the National Academy of Sciences USA*, 112 (2015): 12887–92. doi: 10.1073/pnas.1421881112. Spatial abilities: Kaili Rimfeld et al., 'Phenotypic and Genetic Evidence for a Unifactorial Structure of Spatial Abilities', *Proceedings of the National Academy of Sciences USA*, 114 (2017): 2777–82. doi:

10.1073/pnas.1607883114. For disorders like schizophrenia, published twin heritability estimates are often much higher than those shown in Table 2. These higher estimates use an approach that converts the twin data to a hypothetical continuum of liability rather than using the more conservative approach of relying on actual twin concordance for diagnoses as in Table 2.

p. 7 *'several other common misunderstandings about heritability'*: Following are five common misunderstandings about heritability that I have encountered. An interesting book about heritability, written by a philosopher of science, is Neven Sesardic, *Making Sense of Heritability* (Cambridge University Press, 2005).

Misunderstanding 1: If the heritability of weight is 70 per cent, this means that 70 per cent of your weight is due to genes and the other 30 per cent is due to environment.

Heritability is not about one individual. It's about individual differences in a population and the extent to which inherited DNA differences account for the differences in weight in that population. Even with a heritability of 70 per cent, a particular person's obesity might be caused entirely by environmental circumstances.

Misunderstanding 2: You cannot separate the effects of nature and nurture on weight because both nature and nurture are essential. I collect metaphors implying that you cannot separate the effects of genes and environment. The most common one is the area of a rectangle. One of the many quotes along these lines is from the neuropsychologist Donald O. Hebb, *A Textbook of Psychology* (W. B. Saunders, 1958, p. 129): 'To ask how much heredity contributes to intelligence is like asking how much the width of a field contributes to its area.' In other words, it is not possible to separate the contributions of length and width to the area of a rectangle because area is the product of length and width, that is, the area of a rectangle does not exist without both length and width. The implication is that genes and environments are like this, meaning that you can't separate their effects. However, in a population of rectangles, the variance of areas of the rectangles could be due entirely to length:

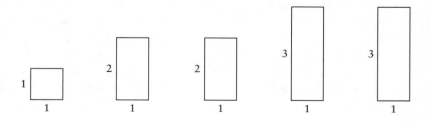

entirely to width:

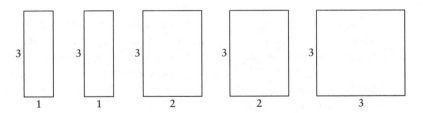

or to both length and width:

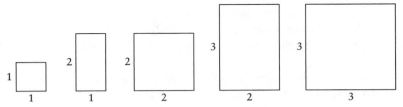

Similarly with weight, the effects of nature and nurture cannot be separated for one individual. Both genes and environment are essential for weight. Without genes there is no individual to weigh, and genes without an environment cannot do anything. The point is that heritability does not refer to one individual but to a population of individuals. Differences between individuals in weight can be due entirely to the environment, entirely to genetics, or to a combination of the two. Heritability is the proportion of variance in weight that can be accounted for by inherited DNA differences.

If the effects of nature and nurture really cannot be separated, this would be just as much an argument against studying environmental influence as against studying genetic influence. It is a sign of reluctance to accept genetic influence that this argument is only applied to studying genetic influence.

Metaphors like the area of a rectangle lead to a related misunderstanding about the word 'interaction'. You multiply length and width to get the area, which means that the effect of length on area depends on width. This metaphor is used to suggest that the effects of nature and nurture interact in the sense that nature depends on nurture. Again, this implies that the effects of nature and nurture on weight cannot be disentangled.

In genetics, interaction means that estimates of genetic effects can differ in different environments. It does not mean that the effects of nature and nurture are inseparable. An example used in the text is that the heritability of weight is higher in wealthier countries where junk food is always at hand than in poorer countries.

Misunderstandings come in when interaction is used to mean that the effects of nature and nurture cannot be separated because the effects of nature depend on nurture. One source of this misunderstanding is that we inherit DNA but the expression of our DNA depends in part on the environment. DNA is not permanently switched on – DNA is expressed as the DNA's product is needed, as described in Chapter 11. Different DNA is expressed in different systems such as brain, heart and liver, even though each cell in all these systems has exactly the same inherited DNA. Within these systems, DNA is turned on and off in response to the environment, from the micro-environment inside the cell to the environment outside the body. You are changing the expression of many neurotransmitter genes in your brain as you read this sentence.

For example, some genes that affect weight are turned on in fat cells and control how much fat you store away in reserve. When there is not much fat in the diet, one particular gene discussed in Chapter 11, *FTO*, is expressed and tells fat cells to stock up on fat. A mutation in the gene makes the *FTO* gene more easily turned on, so more fat is stored. This inherited DNA difference is the single biggest genetic factor in weight, accounting for about a six-pound difference between people with and without this mutation. This gene is switched on in response to food. In our fast-food world with easy access to fatty foods, this inherited DNA difference is doing its thing most of the time. How much fat we consume certainly affects our weight, which counts as an environmental effect. But even with the same diet, this DNA difference in the *FTO* gene would make people differ in weight. The point here is that DNA differences need to be expressed to make a difference but all that we inherit and all that counts for heritability is DNA.

One more related misunderstanding is a version of the phrase 'Man proposes, God disposes.' In this case, the idea is that 'Nature proposes, nurture disposes.' That is, DNA is said to set the limits or potential for development but the environment determines where within those limits an individual ends up. This concept, called *reaction range*, implies that the effects of genes depend on the environment. As shown in the figure with rectangles, this is not the case when we are talking about the origins of individual differences. Genetic effects can occur independently of environmental effects, and vice versa.

This might seem like nit-picking, but it makes an important point about heritability. The 'nature proposes, nurture disposes' notion implies that, although there are potential theoretical limits set by individuals' DNA, their actual development depends on the environment. Heritability is not about potential, what could have been. Instead, it describes the extent to which inherited DNA differences actually create differences between individuals in a population, given the environments in which they live.

Misunderstanding 3: Genetics can't be important for weight because, if you don't eat, you lose weight. Genetic research is about 'what is', not about 'what could be'. People around us differ greatly in weight. If they stopped eating for several days, they would all lose weight. Despite this average weight loss, people would not lose the same amount of weight at the same speed. In starving populations, different genetic factors might affect weight, and heritability might differ from populations with easy access to food.

Heritability is about what causes the differences that we see in a particular population. Many environmental interventions *could* make a difference, but that does not mean that these *are* the factors responsible for variance in weight as it exists in the population. For example, a gastric band placed around the upper section of the stomach restricts the amount of food that can be comfortably eaten. Gastric bands can drastically reduce the body weight of morbidly obese individuals but, obviously, gastric bands have nothing to do with why people are obese in the first place, because gastric bands are surgically inserted. Causes and cures are not necessarily related. Even if the heritability of weight were 100 per cent, gastric bands would still make obese people lose weight.

Nonetheless, knowing 'what is' should be helpful in thinking about 'what could be'. For example, knowing that weight runs in families for reasons of nature, not nurture, means that environmental influences shared by family members, such as diets and lifestyles, do not affect weight. This finding implies that the search for interventions to reduce weight should look for other environmental factors, because these factors currently exist but do not make a difference.

Misunderstanding 4: Genetic influences can't be important because average weight is increasing. Weight has steadily increased over the last fifty years. This increase refers to average differences between groups – we are heavier, on average, than people were fifty years ago. The average change in weight has occurred too quickly to be due to genetic changes, which has wrongly led to the conclusion that genetic factors can't be important.

A remarkable fact is that the heritability of weight has not changed over the decades, despite the substantial increase in average weight. Heritability is about differences between individuals, not average differences between groups. It is an important principle that the causes of average differences between groups are not necessarily related to the causes of individual differences within groups. In the case of weight, individual differences in weight are just as highly heritable now as they were fifty years ago, but the average increase in weight could be entirely environmental in origin. For example, the average increase in weight might be due to greater access to energy-dense foods such as sugar-rich drinks and high-calorie snacks.

This principle also applies to more politically sensitive differences between groups, such as average differences between males and females, between social classes, or between ethnic groups. The causes of average differences are not necessarily related to the causes of individual differences. For example, some of the biggest differences between the sexes are found in childhood psychopathology – boys are many times more likely than girls to be hyperactive or to have autistic symptoms. However, these symptoms are highly heritable for both boys and girls, and genetic studies show that the same genes affect boys and girls. Although DNA differences are substantially responsible for individual differences in these symptoms, they do not appear to account for the average difference between boys and girls. What does account for the average difference? We don't yet know.

Misunderstanding 5: To the extent that genetics is important, there is nothing you can do about it. There is not much you can do about most of the thousands of single-gene disorders. These are disorders caused by a single DNA difference that is necessary and sufficient for the disorder to develop. For example, if people inherit the genetic mutation for Huntington disease, they will die in adulthood from this degenerative neural disorder, regardless of their environment.

For a few single-gene disorders, we can do something about it. One of the rare examples is phenylketonuria (PKU), a single-gene disorder that, if untreated, causes severe intellectual disability. This inherited DNA difference produces a dysfunctional enzyme that cannot break down phenylalanine, one of the essential amino acids that come from certain foods. If a person can't metabolize phenylalanine, it accumulates, and this damages the developing brain. Learning about this inherited metabolic disorder led to a low-tech dietary solution: limit the intake of those foods rich in phenylalanine such as breast milk, eggs and most meats and cheese. The possibility of actually correcting a DNA mutation has been realized recently. A gene-editing technique called CRISPR can efficiently and precisely cut and replace a DNA mutation, as described in Chapter 11.

In contrast, genetic influence on weight and on all psychological traits is not a matter of a hard-wired single-gene mutation. For this reason, gene-editing seems unlikely to be used to alter genes involved in psychological traits. Heritability is the result of thousands of genes of small effect, or *polygenic* genes. The highly polygenic nature of genetic influence is also why heritability does not mean immutability. High heritability for weight implies that these polygenic effects are responsible for weight differences and that existing environmental differences do not make much of a difference.

High heritability of weight means that, on average, across the population, environmental differences such as dietary differences are not a big part of the answer to the question why people differ in weight. Despite this, if you want to lose weight, you can lose weight, but it will be much harder for some people than others because of their genetic propensities. This is another example of the point that heritability is about 'what is', not 'what could be'.

p. 8   *the rate of breast cancer for women who have an identical twin with breast cancer is only 15 per cent'*: Paul Lichtenstein et al., 'Environmental and Heritable Factors in the Causation of Cancer – Analyses of Cohorts of Twins from Sweden, Denmark, and Finland', *New England Journal of Medicine*, 343 (2000): 78–85. doi: 10.1056/NEJM200007133430201.

p. 10   *'People who thought one trait was highly heritable were not the same people who thought the same way about other traits'*: In our 2017 survey of 5,000 young adults in the UK, we found that the average correlation between estimates of heritability across all fourteen traits was only 0.27: Emily Smith-Woolley and Robert Plomin, *Perceptions of Heritability*. Manuscript in preparation.

p. 11   *'20,000 papers published during the past five years alone'*: Ziada Ayorech et al., 'Publication Trends over 55 Years of Behavioral Genetic Research', *Behavior Genetics*, 46 (2016): 603–7. doi: 10.1007 s10519-016-9786-2.

p. 11   *'the first law of behavioural genetics'*: Robert Plomin et al., 'Top 10 Replicated Findings from Behavioral Genetics', *Perspectives on Psychological Science*, 11 (2016): 3–23. doi: 10.1177/1745691615617439.

Chapter 2: How do we know that DNA
makes us who we are?

p. 15 *'the adoption agencies'*: I had to work out the ethical and logistical issues with the adoption agencies. For example, we agreed that the adoption agencies would contact the adoptive parents and ask them to participate in the study only after adoption was agreed so that adoptive parents would feel no pressure to participate. Then I had to get approval from the university's ethical review board. All research at universities needs to be approved and monitored by a formally designated ethics panel to protect the rights and welfare of people participating in research. It was relatively easy for me to get the ethical review board's approval because the major issues of confidentiality and anonymity had already been resolved with the adoption agencies.

p. 17 *'the Colorado Adoption Project . . . continues today, with the children now in their forties'*: Sally-Anne Rhea et al., 'The Colorado Adoption Project', *Twin Research and Human Genetics*, 16 (2013): 358–65. doi: 10.1017/thg.2012.109.

p. 17 *'colorado Adoption Project . . . results have been described in four books and in hundreds of research articles'*: The results described in this section are available in Stephen Petrill et al., *Nature, Nurture, and the Transition to Early Adolescence* (Oxford University Press, 2003).

p. 19 *'social, Genetic and Developmental Psychiatry Centre'*: Peter McGuffin and Robert Plomin, 'A Decade of the Social, Genetic and Developmental Psychiatry Centre at the Institute of Psychiatry', *British Journal of Psychiatry*, 185 (2004): 280–82. doi: 10.1192/bjp.185.4.280.

p. 19 *'a twin study in Colorado that focused on infancy'*: Robert Plomin et al., 'Individual Differences during the Second Year of Life: The MacArthur Longitudinal Twin Study', in John Colombo and Joseph Fagen (eds.), *Individual Differences in Infancy: Reliability, Stability, and Predictability* (Lawrence Erlbaum Associates, 1990): 431–55.

p. 20 *'Twins Early Development Study (TEDS)'*: Claire Haworth et al., 'Twins Early Development Study (TEDS): A Genetically Sensitive Investigation of Cognitive and Behavioral Development from Childhood to Young Adulthood', *Twin Research and Human Genetics*, 16 (2013): 117–25. doi: 10.1017/thg.2012.91.

p. 22 *'TEDS findings have been reported in more than 300 scientific papers and in 30 PhD dissertations'*: Links to these papers can be found on the TEDS website, clicking on 'Research' and then 'Scientific Publications': https://www.teds.ac.uk/.

p. 24 *'the statistics of individual differences in greater detail, using the correlation between weight and height as an example'*: Instead of focusing on averages, the statistics of individual differences focuses on variability. In the TEDS twin study, we assessed weight at the age of sixteen for 2,000 twin pairs. Their average weight is 130 pounds, but they vary in weight from 75 pounds to 250 pounds, as shown in the figure below. The figure shows what is called the normal distribution, the bell-shaped curve, with most scores near the mean and fewer scores as you look towards the low or high extremes. The distribution for weight is not quite normal because the obesity epidemic is responsible for disproportionate numbers of heavier individuals. That is, there is a longer tail on the right side of the distribution.

The distribution of weight in 16-year-olds

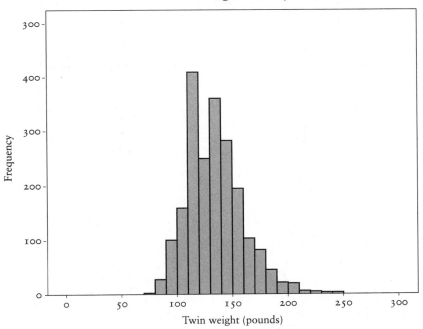

Variance is a statistic that describes this variability, that is, how far individuals' weights are spread out from their mean. It is based on each individual's difference from the mean. An individual who weighs 130 pounds adds nothing to the variance. Someone who weighs 200 pounds adds a lot to the variance. The 200-pounder is 70 pounds above the mean of 130 pounds. This individual adds a lot to the variance, because 70 pounds squared is 4,900.

Covariance is key because it is an index of the strength of the association between two variables. It is called covariance because it indicates the extent to which variance covaries between two variables. As just noted, variance is calculated by squaring each individual's deviation from the average. To calculate covariance, each individual's deviation from the average on one variable is multiplied by the individual's deviation from the average on the other variable. Covariance is the average of these products across individuals. So, covariance will be substantial if people who are well above average on one variable are also well above average on the other variable.

Correlation is the proportion of variance that covaries. It divides the covariance by the variance, which neatly converts covariance to make it more interpretable on a zero-to-one scale. If the two variables covary completely, the covariance equals the variance and the correlation is 1. You can visualize a correlation from a scatter plot. No doubt you have noticed that taller people are heavier. The next figure shows a scatter plot between weight and height from the sixteen-year-olds in my TEDS twin study.

Scatterplot showing the correlation between weight and height in 16-year-olds

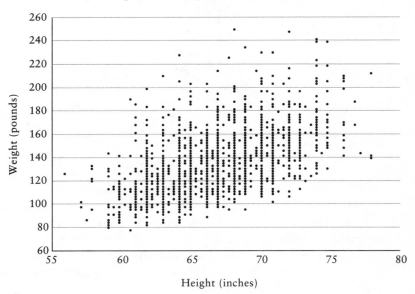

The correlation is 0.6, meaning that 60 per cent of the variance of weight and height covaries. If the correlation were 0, the scatterplot would look round rather than oval, indicating no association between the two variables. If the correlation were 1, the scatterplot would just be a straight line. Scores on weight could perfectly predict height, and vice versa.

The correlation of 0.6 is in between these extremes. The figure clearly shows that heavier people are taller, but there are exceptions. For example, the dot at the top in the centre is one of the heaviest sixteen-year-olds, weighing in at 250 pounds, who is only of average height. Because weight correlates so substantially with height, weight is often adjusted for height to get a purer measure of weight independent of height. One widely used adjustment is called *body mass index*.

p. 26 'the Minnesota Study of Twins Reared Apart': The data on weight come from Thomas J. Bouchard and Matt McGue, 'Familial Studies of Intelligence: A Review', *Science*, 212 (1981): 1055–9. doi: 10.1126/science.7195071. An overview of the study is also available: Nancy L. Segal, *Born Together – Reared Apart* (Harvard University Press, 2012).

p. 26 'Swedish Adoption/Twin Study of Aging': Nancy L. Pedersen et al., 'The Swedish Adoption/Twin Study of Aging: An Update', *Acta Geneticae Medicae et Gemellologiae*, 40 (1991): 7–20. doi: org/10.1017/S0001566600006681.

p. 27 'In TEDS, the MZ correlation for weight is 0.84': This MZ correlation is only slightly greater than the correlation for MZ twins reared apart (0.75). This suggests that twins who spend their whole life together in the same home are only slightly more similar than twins who grew up in different homes. I highlight this finding later, after discussing adoption studies.

p. 27 'the heritability of weight increases from about 40 per cent in early childhood to about 60 per cent in adolescence to about 80 per cent in adulthood': Karri Silventoinen et al., 'Genetic and Environmental Effects on Body Mass Index from Infancy to the Onset of Adulthood: An Individual-based Pooled Analysis of 45 Twin Cohorts Participating in The COllaborative Project of Development of Anthropometrical Measures in Twins (CODATwins) Study', *American Journal of Clinical Nutrition*, 104 (2016): 371–9. doi: 10.3945/ajcn.116.130252.

p. 27 'CAP results for body weight': Robert Plomin et al., *Nature and Nurture during Infancy and Early Childhood* (Cambridge University Press, 1988). doi: 10.1017/CBO9780511527654.

p. 29 *'Putting all of these twin and adoption data together . . . comes up with an estimate of about 70 per cent for heritability of weight'*: One of the important advances in twin and adoption research is called *model-fitting*, which puts all the data together. Model-fitting can simultaneously analyse all of the data from family, twin and adoption studies and come up with a single estimate of heritable influence. It also makes assumptions explicit – such as assumptions about non-additive genetic variance and age changes in genetic effects – and tests the fit of these assumptions. Model-fitting heritability estimates for adult weight are 70 per cent.

What about other measures related to weight, such as body mass index (weight corrected for height), waist circumference and skinfold thickness? Genetic research yields similarly high heritability estimates for these measures. Genetic research using a technique called *multivariate genetic analysis* also reveals that the same genes largely (about 80 per cent) affect these different measures of weight.

p. 29 *'The point is that these very different designs – twin and adoption studies – converge on a simple but powerful conclusion'*: One exception might be self-reported data for personality. We have found that adoption data yield much lower heritability estimates than twin studies, which we attributed to non-additive genetic influence on personality. Robert Plomin et al., 'Adoption Results for Self-reported Personality: Evidence for Non-additive Genetic Effects?', *Journal of Personality and Social Psychology*, 75 (1998): 211–18. doi: 10.1037/0022-3514.75.1.211.

p. 29 *'nuances about differences in twin and adoption designs'*: Finding that twin studies yield the highest heritability estimate, 80 per cent, points to the importance of a particular type of genetic influence detected only in MZ twins. MZ twins are like clones in that their inherited DNA sequence is identical. In contrast, first-degree relatives – siblings, including DZ twins, as well as parents and their children – are not really 50 per cent similar. They are only 50 per cent similar for what is called *additive genetic effects*, effects that 'add up' individually. Because MZ twins have identical DNA, only MZ twins capture non-additive genetic effects, which account for about 10 per cent of the heritability of weight. This is the primary reason why heritability in twin studies is greater than estimates from siblings and parents.

p. 29 *'Does heritability differ in different populations? The answer is "not much"'*: Karri Silventoinen et al., 'Genetic and Environmental Effects on Body Mass Index from Infancy to the Onset of Adulthood: An

Individual-based Pooled Analysis of 45 Twin Cohorts Participating in The COllaborative Project of Development of Anthropometrical Measures in Twins (CODATwins) Study', *American Journal of Clinical Nutrition*, 104 (2016): 371–9. doi: 10.3945/ajcn.116.130252.

p. 29 '*heritability of weight may be higher in richer countries*': J. Min et al., 'Variation in the Heritability of Body Mass Index Based on Diverse Twin Studies: A Systematic Review', *Obesity Research*, 14 (2013): 871–82. doi: 10.1111/obr.12065.

p. 29 '*A recent review of twin studies looked at 18,000 traits in 2,700 publications that included nearly 15 million twin pairs*': Tinca Polderman et al., 'Meta-analysis of the Heritability of Human Traits Based on Fifty Years of Twin Studies', *Nature Genetics*, 47 (2015): 702–9. doi: 10.1038/ng.3285.

p. 30 '*sex differences account for less than 1 per cent of the variance*': Janet S. Hyde, 'Gender Similarities and Differences', *Annual Review of Psychology* (2014). 65: 373–98. doi: 10.1146/annurev-psych-010213-115057.

Chapter 3: The nature of nurture

p. 32 '*In this book I focus on the five most significant findings*': In fact, I and my colleagues have described ten of the biggest findings that have emerged during the past few decades: Robert Plomin et al., 'Top 10 Replicated Findings from Behavioral Genetics', *Perspectives on Psychological Science*, 11 (2016): 3–23. doi: 10.1177/1745691615617439.

p. 32 '*Why Most Published Research Findings are False*': John P. A. Ioannidis, 'Why Most Published Research Findings are False', *PLoS Medicine*, 2 (2005): e124. doi: 10.1371/journal.pmed.0020124.

p. 33 '*Failures to replicate are popping up all over science*':

> *In medicine*: C. Glenn Begley and Lee M. Ellis, 'Raise Standards for Preclinical Cancer Research', *Nature*, 483 (2012): 531–3. doi:10. 1038/483531a.

> *In pharmacology*: Florian Prinz et al., 'Believe It or Not: How Much Can We Rely on Published Data on Potential Drug Targets?', *Nature Reviews Drug Discovery*, 10 (2011): 712. doi:10.1038/ nrd3439-c1.

> *In neuroscience*: Wouter Boekel et al., 'A Purely Confirmatory Replication Study of Structural Brain–Behavior Correlations', *Cortex*, 66 (2015): 115–33. doi:10.1016/j.cortex.2014.11.019. Anders Eklund et al., 'Cluster Failure: Why fMRI Inferences for Spatial Extent Have

Inflated False-positive Rates', *Proceedings of the National Academy of Sciences USA*, 113 (2016): 7900–905. doi: 10.1073/pnas.1602413113.

p. 33 '*In relation to psychology, an influential paper in the journal* Science *reported that more than half of 100 studies in top journals failed to replicate*': Alexander A. Aarts et al., 'Estimating the Reproducibility of Psychological Science', *Science*, 349 (2015). doi: 10.1126/science. aac4716. A critique of this influential paper concluded that the situation in the behavioural sciences was not quite so dire: Daniel T. Gilbert et al., 'Estimating the Reproducibility of Psychological Science', *Science*, 351 (2016). doi: 10.1126/science.aad7243. However, a response to this critique indicates that the jury is still out on the severity of the problem: Christopher J. Anderson et al., 'Response to Comment on "Estimating the Reproducibility of Psychological Science"', *Science*, 351 (2016). doi: 10.1126/science.aad9163.

p. 33 '*As the physicist Richard Feynman said, "The first principle is that you must not fool yourself – and you are the easiest person to fool"*': Richard Feynman, *Surely You're Joking, Mr Feynman* (Vintage, 1992, p. 343).

p. 34 '*how to fix these cracks in the foundation of science*': Stepping back from statistical issues, I believe that what is needed most is to overcome the disconnect between what is good for scientists and what is good for science. What is good for scientists is getting published in good journals. What is good for science is getting it right. Getting it right is much easier to say than to do. However, at the risk of sounding sanctimonious, the real pleasure of science is making new, true discoveries that replicate. Getting it right. Brian A. Nosek et al., 'Restructuring Incentives and Practices to Promote Truth over Publishability', *Perspectives on Psychological Science*, 7 (2012): 615–31. doi: 10.1177/1745691612459058. John P. A. Ioannidis, 'How to Make More Published Research True', *PLoS Medicine*, 11 (2014). e1001747. doi: 10.1371/journal.pmed.1001747.

p. 34 '*Why do findings in behavioural genetics replicate so strongly?*': Robert Plomin et al., 'Top 10 Replicated Findings from Behavioral Genetics', *Perspectives on Psychological Science*, 11 (2016): 3–23. doi: 10.1177/1745691615617439.

p. 34 '*behavioural genetics has been the most controversial topic in psychology during the twentieth century*': Steven Pinker, *The Blank Slate: The Modern Denial of Human Nature* (Penguin, 2003).

p. 35 '*the nature of nurture*': Robert Plomin and Cindy S. Bergeman, 'The Nature of Nurture: Genetic Influence on "Environmental" Measures (with Open Peer Commentary and Response)', *Behavioral and Brain Sciences*, 14 (1991): 373–428. doi:10.1017/S0140525X00070278.

p. 35 *'When research was done to test [Freud's] ideas, little support was found for them'*: Hans Eysenck, *Decline and Fall of the Freudian Empire* (Pelican, 1986). Richard Webster and Malcolm Macmillan, *Freud Evaluated: The Completed Arc* (MIT Press, 1997). Richard Webster, *Why Freud was Wrong: Sin, Science and Psychoanalysis* (The Orwell Press, 2005).

p. 35 *'The philosopher of science Karl Popper claimed that Freud's theories were presented in a form that made them impossible to disprove'*: Karl Popper, *Conjectures and Refutations: The Growth of Scientific Knowledge* (Routledge and Kegan Paul, 1963).

p. 38 *'genetic analysis of stressful life events [using] twins reared apart as well as twins reared together'*: Robert Plomin et al., 'Genetic Influence on Life Events During the Last Half of the Life Span', *Psychology and Aging*, 5 (1990): 25–30. doi: 10.1037/0882-7974.5.1.25.

p. 38 *'the Social Readjustment Rating Scale'*: Thomas H. Holmes and Richard H. Rahe, 'The Social Readjustment Rating Scale', *Journal of Psychosomatic Research*, 11 (1967): 213–18.

p. 39 *'Subsequent research has shown that certain personality traits account for a third of the genetic influence on divorce'*: Victor Jocklin et al., 'Personality and Divorce: A Genetic Analysis', *Journal of Personality and Social Psychology*, 71 (1996): 288–99. http://dx.doi.org/10.1037/0022-3514.71.2.288.

p. 39 *'the link between divorce in parents and divorce in their children is forged genetically, not environmentally'*: Jessica E. Salvatore et al., 'Genetics, the Rearing Environment, and the Intergenerational Transmission of Divorce: A Swedish National Adoption Study', *Psychological Science*, 29 (2018): 370–78. doi: 10.1177/0956797617734864. Epub 18 January 2018.

p. 40 *'more than 2,000 studies exploring [the effect of children's television viewing] on [their] development'*: David Pearl, Lorraine Brouthilet and Joyce B. Lazar, *Television and Behaviour: Ten Years of Scientific Progress and Implications for the Eighties, Volume 1* (US Government Printing Office, 1982).

p. 42 *In 1989 I wrote a paper showing genetic influence on individual differences in children's television viewing*: Robert Plomin et al., 'Individual Differences in Television Viewing in Early Childhood: Nature as Well as Nurture', *Psychological Science*, 1 (1990): 371–7. doi: 10.1111/j.1467-9280.1990.tb00244.x

p. 42 *positive news story about the [genetics of television viewing] in . . . Science*': 'News & Comment: TV Attachment Inherited?', *Science*, 250 (1990): 1335.

p. 42 *Genetic analyses of TV viewing are of uncertain meaning*: Richard J. Rose, 'Genes and Human Behavior', *Annual Review of Psychology*, 46 (1995): 625–54.

p. 44 '*In 1991 I published a paper reviewing the results of these eighteen studies, which I called "The Nature of Nurture"*': Robert Plomin and Cindy S. Bergeman, 'The Nature of Nurture: Genetic Influence on "Environmental" Measures (with Open Peer Commentary and Response)', *Behavioral and Brain Sciences*, 14 (1991): 373–428. doi:10.1017/S0140525X00070278.

p. 44 '*Characteristics of adolescents' peer groups are especially highly heritable*': Beth Manke et al., 'Genetic Contributions to Adolescents' Extrafamilial Social Interactions: Teachers, Best Friends, and Peers', *Social Development*, 4 (1995): 238–56. doi: 10.1111/j.1467-9507.1995.tb00064.x.

p. 45 '*For quality of support, we found that a third of the differences between people could be explained by genetic factors*': Cindy S. Bergeman et al., 'Genetic and Environmental Influences on Social Support: The Swedish Adoption/Twin Study of Aging (SATSA)', *Journal of Gerontology: Psychological Sciences*, 45 (1990): P101–P106. doi: 10.1093/geronj/45.3.P101.

p. 46 '*A recent twin analysis showed that individual differences in the use of Facebook in young adults yielded a heritability of 25 per cent*': Ziada Ayorech et al., 'Personalized Media: A Genetically Sensitive Investigation of Individual Differences in Online Media Use', *PLoS One*, 12 (2017): e0168895. doi: 10.1371/journal.pone.0168895.

p. 48 '*the number of children's scrapes and bruises shows genetic influence*': Kay Philipps and Adam P. Matheny, 'Quantitative Genetic Analysis of Injury Liability in Infants and Toddlers', *American Journal of Medical Genetics. Part B, Neuropsychiatric Genetics*, 60 (1995): 64–71. doi: 10.1002/ajmg.1320600112.

p. 49 '*We showed that about half of [the correlation between the HOME and children's development] can be attributed to genetics*': Robert Plomin, John C. Loehlin and John C. DeFries, 'Genetic and Environmental Components of "Environmental" Influences', *Developmental Psychology*, 21 (1985): 391–402. doi: 10.1037/0012-1649.21.3.391.

p. 50 '*we found that genetics accounts for over half of the correlation [between social support and well-being]*': Cindy S. Bergeman et al.,

'Genetic Mediation of the Relationship between Social Support and Psychological Well-being', *Psychology and Aging*, 6 (1991): 640–46. doi: 10.1037/0882-7974.6.4.640.

p. 50 *'Since 1991, more than a hundred papers have looked at environmental measures in genetically sensitive studies'*: Reut Avinun and Ariel Knafo, 'Parenting as a Reaction Evoked by Children's Genotype: A Meta-analysis of Children-as-Twins Studies', *Personality and Social Psychology*, 18 (2014): 87–102. doi: 10.1177/1088868313498308. Kenneth S. Kendler and Jessica H. Baker, 'Genetic Influences on Measures of the Environment: A Systematic Review', *Psychological Medicine*, 37 (2007): 615–26. doi: 10.1017/S0033291706009524. Ashlea M. Klahr and S. Alexandra Burt, 'Elucidating the Etiology of Individual Differences in Parenting: A Meta-analysis of Behavioral Genetic Research', *Psychological Bulletin*, 140 (2014): 544–86. doi: 10.1037/a0034205.

p. 51 *Genetic influence on environmental measures has been found most recently in DNA studies*: Eva Krapohl et al., 'Widespread Covariation of Early Environmental Exposures and Trait-associated Polygenic Variation', *Proceedings of the National Academy of Sciences USA*, 114 (2017): 11727–32. doi: 10.1073/pnas.1707178114. Robert Plomin, 'Genotype–Environment Correlation in the Era of DNA', *Behavior Genetics*, 44 (2014): 629–38. doi: 10.1007/s10519-014-9673-7

Chapter 4: DNA matters more as time goes by

p. 52 *'The construct of intelligence captures what diverse cognitive tests have in common, which is why intelligence is often referred to as gen-*eral cognitive ability *or* g*'*: Originally defined by Charles Spearman, 'General Intelligence, Objectively Determined and Measured', *American Journal of Psychology*, 15 (1904): 201–92. Among dozens of books on intelligence, an especially readable recent one is by Stuart Ritchie, *Intelligence: All That Matters* (Hodder & Stoughton, 2015).

p. 53 *'According to the majority view of intelligence researchers, the core of intelligence is'*: Linda S. Gottfredson, 'Mainstream Science on Intelligence: An Editorial with 52 Signatories, History, and Bibliography', *Intelligence*, 24 (1994): 13–23. doi: 10.1016/S0160-2896(97)90011-8.

p. 53 *Intelligence involves not specific modules but brain processes working in concert to solve problems*: Ian J. Deary et al., 'The Neuroscience of

Human Intelligence Differences', *Nature Reviews Neuroscience*, 11 (2010): 201–11. doi: 10.1038/nrn2793.

p. 53 *'Intelligence is one of the best predictors of key outcomes'*: Linda S. Gottfredson, 'Why g Matters: The Complexity of Everyday Life', *Intelligence*, 24 (1997): 79–131. Frank L. Schmidt and John E. Hunter, 'The Validity and Utility of Selection Methods in Personnel Psychology: Practical and Theoretical Implications of 85 Years of Research Findings', *Psychological Bulletin*, 124 (1998): 262–74. doi: 10.1037/0033-2909.124.2.262.

p. 53 *'During the past century, genetic research on intelligence was in the eye of the storm of the nature–nurture debate in the social sciences'*: Editorial, 'Intelligence Research Should Not be Held Back by Its Past', *Nature*, 545 (25 May 2017): 385–6.

p. 53 *'Soviet hypothesis that heritability disappears after childhood'*: Robert Plomin, 'Foreword' in Yulia Kovas et al. (eds.), *Behavioural Genetics for Education* (Palgrave Macmillan UK, 2016). doi: 10.1057/9781137437327.

p. 54 *'The Louisville Twin Study first suggested that heritability increases for intelligence during infancy and childhood'*: Ronald S. Wilson, 'The Louisville Twin Study: Developmental Synchronies in Behavior', *Child Development*, 54 (1983): 298–316. doi: 10.1111/j.1467-8624.1983.tb03874.x.

p. 54 *'a dramatic confirmation of this finding [of increasing heritability for intelligence] came from our Colorado Adoption Project'*: Robert Plomin et al., 'Nature, Nurture, and Cognitive Development from 1 to 16 Years: A Parent–Offspring Adoption Study', *Psychological Science*, 8 (1997): 442–7. doi: 10.1111/j.1467-9280.1997.tb00458.x.

p. 54 *'Further support for the hypothesis of increasing heritability came from a 2010 consortium of twin studies'*: Claire M. A. Haworth et al., 'The Heritability of General Cognitive Ability Increases Linearly from Childhood to Young Adulthood', *Molecular Psychiatry*, 15 (2010): 1112–20. doi: 10.1038/mp.2009.55.

p. 54 *'in 2013, a meta-analysis brought together results from all twin and adoption studies of intelligence and confirmed the developmental increase in heritability'*: Daniel A. Briley and Elliot M. Tucker, 'Explaining the Increasing Heritability of Cognitive Ability across Development: A Meta-analysis of Longitudinal Twin and Adoption Studies', *Psychological Science*, 24 (2013): 1704–13. doi: 10.1177/0956797613478618.

p. 55 *'The few available studies of later life suggest that increasing heritability continues throughout adulthood to about 80 per cent heritability at the age of sixty-five'*: Matt McGue and Kaare Christensen, 'Growing Old but

Not Growing Apart: Twin Similarity in the Latter Half of the Lifespan',
Behavior Genetics, 43 (2013): 1–12. doi: 10.1007/s10519-012-9559-5.

p. 55 *'other traits show . . . little developmental change in heritability, most
notably personality and school achievement'*: Eric Turkheimer et al.,
'A Phenotypic Null Hypothesis for the Genetics of Personality', *Annual
Review of Psychology*, 65 (2014): 514–40. doi: 10.1146/annurev-psych-
113011-143752. Yulia Kovas et al., 'Literacy and Numeracy are More
Heritable than Intelligence in Primary School', *Psychological Science*,
24 (2013): 2048–56. doi: 10.1177/0956797613486982.

p. 55 *'School achievement is more heritable than intelligence in the early
school years'*: Yulia Kovas et al., 'Literacy and Numeracy are More
Heritable than Intelligence in Primary School', *Psychological Science*,
24 (2013): 2048–56. doi: 10.1177/0956797613486982.

p. 56 *'male pattern baldness is highly heritable but the effects of these genes
do not show up until hormones change in mid-life'*: Saskia P. Hage-
naars et al., 'Genetic Prediction of Male Pattern Baldness', *PLoS
Genetics*, 13 (2017): e1006594. doi: 10.1371/journal.pgen.1006594.

p. 56 *'genetic effects on intelligence in Year 2 correlate 0.7 with genetic effects
on intelligence in Year 4'*: Yulia Kovas et al., 'The Genetic and Environmental Origins of Learning Abilities and Disabilities in the Early School
Years', *Monographs of the Society for Research in Child Development*,
72 (2007): 1–144. doi: 10.1111/j.1540-5834.2007.00453.x.

p. 56 *'Genetic correlations from age to age are even greater after childhood'*: Daniel A. Briley and Elliot M. Tucker, 'Explaining the
Increasing Heritability of Cognitive Ability across Development: A
Meta-analysis of Longitudinal Twin and Adoption Studies', *Psychological Science*, 24 (2013): 1704–13. doi: 10.1177/0956797613478618.

p. 56 *'A recent DNA study strongly supports these results from twin studies, finding 90 per cent overlap in the genes that affect intelligence in
childhood and adulthood'*: Suzanne Sniekers, 'Genome-wide Association Meta-analysis of 78,308 Individuals Identifies New Loci and
Genes Influencing Human Intelligence', *Nature Genetics*, 49 (2017):
1107–12. doi: 10.1038/ng.3869.

p. 57 *'the same genetic factors snowball into larger and larger effects, a process that is known as* genetic amplification': Robert Plomin and John
C. DeFries, 'A Parent–Offspring Adoption Study of Cognitive Abilities
in Early Childhood', *Intelligence*, 9 (1985): 341–56. doi: 10.1016/0160-
2896(85)90019-4. Recent model-fitting meta-analyses have confirmed
genetic amplification: Daniel A. Briley and Elliot M. Tucker-Drob,
'Explaining the Increasing Heritability of Cognitive Ability across

Development: A Meta-analysis of Longitudinal Twin and Adoption Studies', *Psychological Science*, 24 (2013): 1704–13. doi: 10.1177/0956797613478618.

Chapter 5: Abnormal is normal

p. 58 *'Fifty per cent of us will have a diagnosable psychological problem in our lifetime and 20 per cent will have had one within the last year'*: Zachary Steel et al., 'The Global Prevalence of Common Mental Disorders: A Systematic Review and Meta-analysis 1980–2013', *International Journal of Epidemiology*, 43 (2014): 476–93. doi: 10.1093/ije/dyu038.

p. 59 *'the same genes are responsible for reading disability and . . . ability'*: Robert Plomin and Yulia Kovas, 'Generalist Genes and Learning Disabilities', *Psychological Bulletin*, 131 (2005): 592–617. doi: 10.1037/0033-2909.131.4.592. Similar results have been found for the high and low extremes of other psychological traits, for example, for low and high intelligence – they are quantitatively, not qualitatively, different genetically from the rest of the distribution of intelligence: Robert Plomin et al., 'Common Disorders are Quantitative Traits', *Nature Reviews Genetics*, 10 (2009): 872–8. doi: 10.1038/nrg2670. One exception is severe intellectual disability, which is genetically distinct from the rest of the distribution of intelligence and affected by rare mutations with large effects: Avi Reichenberg et al., 'Discontinuity in the Genetic and Environmental Causes of the Intellectual Disability Spectrum', *Proceedings of the National Academy of Sciences USA*, 113 (2016): 1098–1103. doi: 10.1073/pnas.1508093112.

p. 60 *'Genes originally identified because they are associated with a common disorder turn out to be associated with normal variation throughout the distribution'*: Saskia Selzam et al., 'Genome-wide Polygenic Scores Predict Reading Performance throughout the School Years', *Scientific Studies of Reading*, 21 (2017): 334–9. doi: 10.1080/10888438.2017.1299152.

p. 60 *OGOD hypothesis (one gene, one disorder)*: Robert Plomin et al., 'The Genetic Basis of Complex Human Behaviors', *Science*, 264 (1994): 1733–9. doi: 10.1126/science.8209254.

p. 62 *'This genetic research leads to a momentous conclusion: There are no qualitative disorders, only quantitative dimensions'*: Robert Plomin et al., 'Common Disorders are Quantitative Traits', *Nature Reviews Genetics*, 10 (2009): 872–8. doi: 10.1038/nrg2670.

p. 62 *'This view of what we call abnormal as part of the normal distribution of differences is already changing the way we think about mental health and illness'*: Thomas Insel et al., 'Research Domain Criteria (RDoC): Toward a New Classification Framework for Research on Mental Disorders', *American Journal of Psychiatry*, 167 (2010): 748–51. doi: 10.1176/appi.ajp.2010.09091379.

p. 63 *'Who has not sometimes experienced some of these symptoms?'*: Louise C. Johns et al., 'The Continuity of Psychotic Experiences in the General Population', *Clinical Psychology Review*, 21 (2001): 1125–41. doi: 10.1016/S0272-7358(01)00103-9.

Chapter 6: Generalist genes

p. 66 *'I call this topic generalist genes'*: Robert Plomin and Yulia Kovas, 'Generalist Genes and Learning Disabilities', *Psychological Bulletin*, 131 (2005): 592–617. doi: 10.1037/0033-2909.131.4.592.

p. 66 *'Family studies first suggested that genetic effects might be general across disorders rather than specific to each disorder'*: Kaite A. McLaughlin et al., 'Parent Psychopathology and Offspring Mental Disorders: Results from the WHO World Mental Health Surveys', *British Journal of Psychiatry*, 200 (2012): 290–99. doi: 10.1192/bjp.bp.111.101253.

p. 67 *'generalized anxiety disorder and major depressive disorder are the same thing genetically'*: Kenneth S. Kendler et al., 'Major Depression and Generalized Anxiety Disorder – Same Genes, (Partly) Different Environments', *Archives of General Psychiatry*, 49 (1992): 716–22.

p. 67 *'The same result emerged from two dozen twin studies that looked at the genetic overlap between dimensions of anxiety symptoms and dimensions of depression symptoms'*: Christel M. Middeldorp et al., 'The Co-morbidity of Anxiety and Depression in the Perspective of Genetic Epidemiology. A Review of Twin and Family Studies', *Psychological Medicine*, 35 (2005): 611–24. doi: 10.1017/S003329170400412X.

p. 67 *These three genetic clusters also overlap to create a general genetic factor of psychopathology*: Avshalom Caspi and Terrie Moffitt, 'All for One and One for All: Mental Disorders in One Dimension', *American Journal of Psychiatry*. Advance online publication. doi: 10.1176/appi.ajp.2018.17121383.

p. 67 *'Not all genetic effects are general'*: One interesting exception to the rule of generalist genes is autism. Three clusters of symptoms are used to diagnose what is now officially called autism spectrum disorder – social impairment, communication difficulties and rigid and repetitive behaviours. Diagnosis of autism requires impairments in each of these areas. If you ask why this is the case, there is no answer other than historical happenstance. Genetic research shows that most of the genes that affect one of these areas are different from the genes that affect the other areas. Putting three genetically unrelated things together is likely to be the reason why research trying to find genes for autism has been less successful than for other disorders. From a genetic perspective, autism diagnosed as a triad of symptoms does not exist. Each of the three areas is a real problem, but they need to be studied separately because they are different genetically. Francesca Happé et al., 'Time to Give Up on a Single Explanation for Autism', *Nature Neuroscience*, 9 (2006): 1218–20. doi: 10.1038/nn1770.

p. 68 *'most DNA differences found to be associated with schizophrenia also showed associations with bipolar disorder'*: David H. Kavanagh et al., 'Schizophrenia Genetics: Emerging Themes for a Complex Disorder', *Molecular Psychiatry*, 20 (2015): 72–6. doi: 10.1038/mp.2014.148. Hong Lee et al., 'Genetic Relationship between Five Psychiatric Disorders Estimated from Genome-wide SNPS', *Nature Genetics*, 45 (2013): 984–94. doi: 10.1038/ng.2711.

p. 68 *'genetic effects are also general across cognitive abilities'*: Robert Plomin and Yulia Kovas, 'Generalist Genes and Learning Disabilities', *Psychological Bulletin*, 131 (2005): 592–617. doi: 10.1037/0033-2909.131.4.592.

p. 68 *'Education-related skills such as reading, mathematics and science show even higher genetic correlations'*: Yulia Kovas et al., 'The Genetic and Environmental Origins of Learning Abilities and Disabilities in the Early School Years', *Monographs of the Society for Research in Child Development*, 72 (2007): 1–144. doi: 10.1111/j.1540-5834.2007.00453.x.

p. 69 *'The genetic correlation between reading familiar words and non-words is 0.9'*: Nicole Harlaar et al., 'Correspondence between Telephone Testing and Teacher Assessments of Reading Skills in a Sample of 7-year-old Twins: II. Strong Genetic Overlap', *Reading and Writing: An Interdisciplinary Journal*, 18 (2005): 401–23. doi: 10.1007/s11145-005-0271-1.

p. 69 *'Genetic correlations among the dozen spatial tests were on average greater than 0.8'*: Kaili Rimfeld et al., 'Phenotypic and Genetic Evidence for a Unifactorial Structure of Spatial Abilities', *Proceedings of the National Academy of Sciences USA*, 114 (2017): 2777–82. doi: 10.1073/

pnas.1607883114. Nicholas Shakeshaft et al., 'Rotation is Visualisation, 3D is 2D: Using a Novel Measure to Investigate the Genetics of Spatial Ability', *Scientific Reports*, 6 (2016): 30545. doi: 10.1038/srep30545.

p. 70 *'The generalist-genes model makes more sense genetically and evolutionarily than the traditional modularity model'*: Yulia Kovas and Robert Plomin, 'Generalist Genes: Implications for the Cognitive Sciences', *Trends in Cognitive Science*, 10 (2006): 198–203. doi: 10.1016/j.tics.2006.03.001.

p. 70 *'neuroscientists seldom consider individual differences'*: Ian J. Deary et al., 'The Neuroscience of Human Intelligence Differences', *Nature Reviews Neuroscience*, 11 (2010): 201–11. doi: 10.1038/nrn2793.

Chapter 7: Why children in the same family are so different

p. 74 *'genetic study of agreeableness'*: The first genetic study of agreeableness found evidence for shared environment: Cindy S. Bergeman et al., 'Genetic and Environmental Effects on Openness to Experience, Agreeableness, and Conscientiousness: An Adoption/Twin Study', *Journal of Personality*, 61 (1993): 159–79. doi: 10.1111/j.1467-6494.1993.tb01030.x. However, subsequent research has not confirmed this finding: Valerie S. Knopik et al., *Behavioral Genetics, 7th edition* (Worth, 2017).

p. 74 *Grit is another personality trait thought to be due to shared environment, but genetic research shows no influence of shared environmental influence*: Kaili Rimfield et al., 'True Grit and Genetics: Predicting Academic Achievement from Personality', *Journal of Personality and Social Psychology*, 111 (2016): 780–89. doi: 10.1037/pspp0000089.

p. 74 *'experiences shared by family members have no effect on individual differences'*: Valerie S. Knopik et al., *Behavioral Genetics, 7th edition* (Worth, 2017). Tinca Polderman et al., 'Meta-analysis of the Heritability of Human Traits Based on Fifty Years of Twin Studies', *Nature Genetics*, 47 (2015): 702–9. doi: 10.1038/ng.3285.

p. 74 *'Non-shared environment is the name I gave to this mysterious type of environmental influence'*: David C. Rowe and Robert Plomin, 'The Importance of Non-shared Environmental Influences in Behavioral Development', *Developmental Psychology*, 17: 517–31 (1981). doi: 10.1037/0012-1649.17.5.517.

p. 75 *'This finding about the importance of non-shared environment'*: The finding was ignored when it was first noted in 1976 in relation to personality: John C. Loehlin and Robert C. Nichols, *Heredity*,

Environment and Personality (University of Texas, 1976). It was controversial in 1987, when I first reviewed genetic research pointing to this phenomenon: Robert Plomin and Denise Daniels, 'Why are Children in the Same Family So Different from Each Other?', *Behavioral and Brain Sciences*, 10 (1987): 1–16. doi: 10.1017/S0140525X00055941. The controversy was renewed in 1998 when a popular book tackled the topic: Judith R. Harris, *The Nurture Assumption: Why Children Turn Out the Way They Do* (The Free Press, 1988).

p. 75 *'attention has switched to finding any shared environmental influence at all'*: Robert Plomin, 'Commentary: Why are Children in the Same Family So Different? Non-shared Environment Three Decades Later', *International Journal of Epidemiology*, 40 (1991): 582–92. doi: 10.1093/ije/dyq144.

p. 75 *'Intelligence appears to be a major exception to the rule that environmental factors that affect psychological traits are non-shared'*: Valerie S. Knopik et al., *Behavioral Genetics, 7th edition* (Worth, 2017).

p. 75 *'in 1978 a study of adoptive siblings reported a correlation of 0 for adoptive siblings'*: Sandra Scarr and Richard Weinberg, 'The Influence of "Family Background" on Intellectual Attainment', *American Sociological Review*, 43 (1978): 674–92.

p. 75 *'Subsequent studies of older adoptive siblings have found similarly low correlations for intelligence'*: Matt McGue et al., 'Behavioral Genetics of Cognitive Ability: A Life-span Perspective', in Robert Plomin and Gerald E. McClearn (eds.), *Nature, Nurture and Psychology* (American Psychological Association, 1993). 59–76.

pp. 75–6 *'The most impressive evidence comes from a ten-year longitudinal follow-up study of adoptive siblings'*: John C. Loehlin et al., 'Modeling IQ change: Evidence from the Texas Adoption Project', *Child Development*, 60 (1989): 993–1004.

p. 76 *'These findings, supported by twin studies, suggest that shared environment affects intelligence during childhood when children are living at home'*: Claire M. A. Haworth et al., 'The Heritability of General Cognitive Ability Increases Linearly from Childhood to Young Adulthood', *Molecular Psychiatry*, 15 (2010): 1112–20. doi: 10.1038/mp.2009.55.

p. 76 *'School achievement is another apparent exception to the rule [that environmental influence is non-shared]'*: Yulia Kovas et al., 'Literacy and Numeracy are More Heritable than Intelligence in Primary School', *Psychological Science*, 24 (2013): 2048–56. doi: 10.1177/0956797613486982.

p. 76 *'Does the effect of shared environment on school achievement diminish after adolescence, as it does for intelligence?'*: Kaili Rimfeld et al.,

'Genetics Affects Choice of Academic Subjects as Well as Achievement', *Scientific Reports*, 6 (2016): 26373. doi: 10.1038/srep26373.

p. 76 *'The only other exceptions from the hundreds of traits that have been investigated are some religious and political beliefs, for which shared environment accounts for about 20 per cent of the variance'*: Peter K. Hatemi et al., 'Genetic Influences on Political Ideologies: Twin Analyses of 19 Measures of Political Ideologies from Five Democracies and Genome-wide Findings from Three Populations', *Behavior Genetics*, 44 (2014): 282–94. doi: 10.1007/s10519-014-9648-8.

p. 77 *'It is almost as if siblings are living in different families'*: Judith Dunn and Robert Plomin, *Separate Lives: Why Siblings are So Different* (Basic Books, 1990).

p. 77 *'my colleagues David Reiss and Mavis Hetherington and I conducted a ten-year longitudinal study of 700 families with adolescent siblings called "Non-shared Environment in Adolescent Development" (NEAD)'*: David Reiss et al., *The Relationship Code: Deciphering Genetic and Social Patterns in Adolescent Development* (Harvard University Press, 2000).

p. 78 *'The first report of this phenomenon came from NEAD, showing that genetics was largely responsible for the association between differences in parental negativity towards their children and the children's differences in their likelihood of becoming depressed or engaging in antisocial behaviour'*: Alison Pike et al., 'Family Environment and Adolescent Depressive Symptoms and Antisocial Behavior: A Multivariate Genetic Analysis', *Developmental Psychology*, 32 (1996): 590–603. doi: 10.1037/0012-1649.32.4.590.

p. 78 *'most of the non-shared associations that make it to this third step involve the "dark side" of development, such as negative parenting and negative outcomes like depression and antisocial behaviour'*: Bonamy Oliver et al., 'Genetics of Parenting: The Power of the Dark Side', *Developmental Psychology*, 50 (2014): 1233–40. doi: 10.1037/a0035388.

p. 78 *'In desperation, we conducted several studies of identical twins who differed the most in certain traits, for example, in school achievement'*: Kathryn Asbury et al., 'Non-shared Environmental Influences on Academic Achievement at Age 16: A Qualitative Hypothesis-generating Monozygotic-twin Differences Study', *AERA Open*, 2 (2016): 1–12. doi: 10.1177/2332858416673596.

p. 79 *'It is fascinating how often biographies and autobiographies point to chance, such as childhood illness, as critical in explaining why siblings are so different'*: Judith Dunn and Robert Plomin, *Separate Lives: Why Siblings are So Different* (Basic Books, 1990).

p. 79 *'In one of his few jokes, Darwin wrote that during the voyage Fitz-Roy became convinced that "my nose had spoken falsely"'*: Charles Darwin, *The Autobiography of Charles Darwin and Selected Letters*, edited by Francis Darwin (Dover, 1892).

p. 80 *'Sounding like a fortune cookie, Galton went on to say that "tangled strings variously twitched, soon get themselves into tight knots"'*: Francis Galton, *Natural Inheritance* (Macmillan, 1889).

p. 80 *'Identical twin differences for psychological traits, which can only be due to non-shared environment, are not stable across time'*: S. Alexandra Burt and Kelly L. Klump, 'Do Non-shared Environmental Influences Persist over Time? An Examination of Days and Minutes', *Behavior Genetics*, 45 (2015): 24–34. doi: 10.1007/s10519-014-9682-6. Elliot M. Tucker-Drob and Daniel A. Briley, 'Continuity of Genetic and Environmental Influences on Cognition across the Life Span: A Meta-analysis of Longitudinal Twin and Adoption Studies', *Psychological Bulletin*, 140 (2014): 949–79. doi: 10.1037/a0035893.

p. 80 *'In 1987, I wrote about this as the "gloomy prospect" – the possibility that "the salient environment might be unsystematic, idiosyncratic, or serendipitous events"'*: Robert Plomin and Denise Daniels, 'Why are Children in the Same Family So Different from Each Other?', *Behavioral and Brain Sciences*, 10 (1987): 1–16. doi: 10.1017/S0140525X00055941.

Chapter 8: The DNA blueprint

p. 81 *'The Phonics Screening Check ... was among the most highly heritable traits ever reported at this age, with heritabilities of about 70 per cent'*: Nicole Harlaar et al., 'Genetic Influences on Early Word Recognition Abilities and Disabilities: A Study of 7-year-old Twins', *Journal of Child Psychology and Psychiatry*, 46 (2005): 373–84. doi: 10.1111/j.1469-7610.2004.00358.x.

p. 82 *'Education is the field that has been slowest to absorb the messages from genetic research'*: Kathryn Asbury and Robert Plomin, *G is for Genes: What Genetics Can Teach Us about How We Teach Our Children* (Wiley-Blackwell, 2013). doi: 10.1002/9781118482766.

p. 85 *Parents are not carpenters or gardeners*: The developmental psychologist Alison Gopnik comes to a similar view that parents are not carpenters who construct a child. Although caring for children is crucial, parenting is not a matter of shaping them to turn out a particular way. She suggests that parents are more like gardeners, providing conditions for their children to thrive. My view is that parents are not even gardeners, if that implies nurturing and pruning plants to achieve a certain result. The conclusion I reach from the genetic research reviewed in previous chapters is that parents have little systematic effect on their children's outcomes beyond the blueprint that their genes provide. In addition, parents are neither carpenters nor gardeners in the sense that parenting is not a means to an end. It is a relationship and, like our relationships with our partner and friends, our relationship with our children should be based on being with them, not changing them. Alison Gopnik, *The Gardener and the Carpenter: What the New Science of Child Development Tells Us about the Relationship between Parents and Children* (Bodley Head, 2016).

p. 86 *'parents should relax and enjoy their relationship with their children without feeling a need to mould them'*: Anthropologists Robert and Sarah LeVine draw similar conclusions from their studies of parenting practices around the world. Despite great differences in parenting, children turn out to be well-adjusted adults. Robert Levine and Sarah LeVine, *Do Parents Matter?: Why Japanese Babies Sleep Soundly, Mexican Siblings Don't Fight, and Parents Should Just Relax* (Souvenir Press, 2016).

p. 87 *'Ofsted ratings of school quality explained less than 2 per cent of the variance in GCSE scores after correcting for students' achievement in primary school'*: Emily Smith-Woolley et al., 'Ofsted Secondary School Quality is Poor Predictor of Student Academic Achievement and Wellbeing'. Manuscript submitted for publication (2018).

p. 89 *'education is not just preparation for life – education is a big chunk of life itself'*: This is a paraphrase of an idea described by John Dewey, 'My Pedagogic Creed', *School Journal*, 54 (1897): 77–80.

Chapter 9: Equal opportunity and meritocracy

p. 93 *'Are genetic castes inevitable?'*: This question has been bound up in the topic of meritocracy, beginning with sociologist Michael Young's *The Rise and Fall of the Meritocracy* in 1958 (Transaction Publishers). The book was meant as a cautionary tale about the dangers of meritocracy. The rise of meritocracy rests on replacing aristocracy and inherited wealth with talent. The fall of meritocracy is a revolt by the have-nots against the elites, which is eerily like the populist revolt against experts and elites that we see today. These questions reached fever pitch in 1994 with *The Bell Curve: Intelligence and Class Structure in American Life* (The Free Press, 1994) by psychologist Richard J. Herrnstein and political scientist Charles Murray, who warned that society was becoming stratified into an hereditary elite and an underclass. Twenty years later, concerns about meritocracy are on the rise again.

p. 97 *'increased heritability and decreased shared environmental influence after the Second World War, as equality of educational opportunity increased'*: Andrew C. Heath et al., 'Education Policy and the Heritability of Educational Attainment', *Nature*, 314 (1985): 734–6.doi: 10.1038/314734a0. Amelia R. Branigan et al., 'Variation in the Heritability of Educational Attainment: An International Meta-analysis', *Social Forces*, 92 (2013): 109–140. doi: 10.1093/sf/sot076. Dalton Conley and Jason Fletcher, *The Genome Factor* (Princeton University Press, 2017).

p. 97 *'Greater inequality in the US in the twenty-first century'*: François Nielsen and J. Micah Roos, 'Genetics of Educational Attainment and the Persistence of Privilege at the Turn of the 21st Century', *Social Forces*, 94 (2015): 535–61. doi: 10.1093/sf/sov080.

p. 98 *'selective schools do not improve students' [GCSE] achievement once we take into account the fact that these schools preselect students with the best chance of success'*: Emily Smith-Woolley et al., 'Differences in Exam Performance between Pupils Attending Selective and Non-Selective Schools Mirror the Genetic Differences between Them', *NPJ Science of Learning* (2018). Advance online publication. doi: 10.1038/s41539-018-0019-8.

p. 99 *'we have found that this measure of "progress" is still substantially heritable (40 per cent), which means that it is not a pure index of students' "progress" or schools' added value'*: Emily Smith-Woolley and Robert Plomin, 'In the School or in the Genes? The Genetics of Academic Progress'. Manuscript in preparation.

p. 99 *'There may be benefits of grammar and private schools in terms of other outcomes'*: It is difficult to find solid evidence for this, but it is widely accepted that students from private schools dominate the top professions: https://www.suttontrust.com/newsarchive/john-claughton-sees-independent-schools-as-part-of-the-solution-on-social-mobility/.

p. 100 *'students from selective secondary schools are much more likely to be accepted by the best universities, but this benefit largely disappears after controlling for selection factors'*: Emily Smith-Wooley and Robert Plomin, 'Do Selective Secondary Schools Make a Difference at University?' Manuscript in preparation.

p. 100 *'Both occupational status and income are substantially heritable, about 40 per cent in more than a dozen twin studies in developed countries'*: Amelia R. Branigan et al., 'Variation in the Heritability of Educational Attainment: An International Meta-analysis', *Social Forces*, 92 (2013): 109–40. doi: 10.1093/sf/sot076. Dalton Conley and Jason Fletcher, *The Genome Factor* (Princeton University Press, 2017).

p. 102 *'genetic castes, as happened in India, where for thousands of years mating was limited to members of the same caste'*: Analabha Basu et al., 'Genomic Reconstruction of the History of Extant Populations of India Reveals Five Distinct Ancestral Components and a Complex Structure', *Proceedings of the National Academy of Sciences of the United States of America*, 113 (2016): 1594–9. doi: 10.1073/pnas.1513197113.

p. 103 *'As long as downward social mobility as well as upward social mobility occurs, we do not need to fear that genetics will lead to a rigid caste system'*: Although I conclude that genetic castes are not inevitable, many scholars would disagree, most notably Charles Murray and Richard Herrnstein (*The Bell Curve*, The Free Press, 1994). The economist Gregory Clark (*The Son Also Rises: Surnames and the History of Social Mobility*, Princeton University Press, 2014) concludes that social mobility is much lower across centuries and across countries than has been assumed. However, Clark's research relies on analyses of surnames and shows that the social status of families persists for many generations. I think his findings are based on the average success of families, which shows greater persistence over the generations, as compared to individuals within families. Finally, sociologists Dalton Conley and Jason Fletcher (*The Genome Factor*, Princeton University Press, 2017) argue that we are moving towards a 'genotocracy'. This trend is accelerated by assortative mating, the tendency for like-minded individuals to mate.

p. 104 *'My value system suggests that we need to replace meritocracy with a just society'*: This is the theme of a 2016 book, *The Myth of Meritocracy* by James Bloodworth (Biteback Publishing, 2016). On the last page Bloodworth writes: 'Should those who inherit low ability be condemned to a bleak and wretched life based on what is, in essence, the mere lottery of genetics? A more egalitarian society would ensure that everyone could live well, whereas a meritocratic society would endlessly remind the drudges of their worthlessness. A just society is thus not a meritocratic one.'

pp. 104–5 *'60 per cent of the increase in US national income in the last three decades went to just the top 1 per cent of earners'*: This is the most quoted statistic from Thomas Piketty's *Capital in the Twenty-first Century* (Harvard University Press, 2014).

PART TWO: THE DNA REVOLUTION
Chapter 10: DNA: The basics

pp. 109–12 *'Watson and Crick, the most important paper in biology'*: James D. Watson and Francis H. C. Crick, 'Genetical Implications of the Structure of Deoxyribonucleic Acid', *Nature*, 171 (1953): 964–7. The quote 'It has not escaped our notice that the specific pairing we have postulated immediately suggests a possible copying mechanism for the genetic material' is on p. 965.

p. 112 *'we begin life as a single cell and end up with 50 trillions of cells, each with the same DNA'*: This estimate refers only to our own cells. Amazingly, we have at least as many non-human cells living in us as human cells. This is the microbiota of bacteria, fungi, archaea and viruses.

p. 112 *'siblings are, on average, 50 per cent similar'*: Actually, siblings never inherit exactly the same chromosome. When eggs or sperm are formed, members of each chromosome pair make contact and exchange pieces of DNA. This shuffling process creates hybrid chromosomes, a process called *recombination*. For this reason, each egg and each sperm has different recombined chromosomes, which means that siblings cannot inherit exactly the same chromosome. The exception is identical twins, who have exactly the same chromosomes because they come from the same fertilized egg. Despite recombination, siblings are still about 50 per cent similar on average for any particular stretch of DNA, whether it is recombined or not. This is why siblings are similar but also different for psychological traits and why identical twins are more similar than other siblings.

p. 113 *'There may be as many as 80 million SNPs in the world'*: 1,000 Genomes Project Consortium et al., 'An Integrated Map of Genetic Variation from 1,092 Human Genomes', *Nature*, 491 (2012): 56–65. doi: 10.1038/nature11632. David M. Altshuler et al., 'A Global Reference for Human Genetic Variation', *Nature*, 526 (2015): 68–74. doi: 10.1038/nature15393.

p. 113 *'DNA sequence is transcribed by a messenger molecule called RNA'*: We used to think that this RNA message was always translated into amino-acid sequences, which are the building blocks of all proteins. However, DNA transcribed into RNA and translated into amino-acid sequences accounts for only 2 per cent of all DNA. These are the 20,000 classical genes mentioned earlier. Is the other 98 per cent of DNA junk? We now know that as much as half of all DNA cannot be junk, because it is transcribed into RNA even though it is not translated into RNA. Instead of being called junk DNA, it is called non-coding DNA because it does something, even though it does not code for amino-acid sequences. One reason why it must be important is that at least 10 per cent of this non-coding DNA is the same across related species, suggesting that it has some adaptive function because it has been conserved evolutionarily. Other more direct research suggests that as much as 80 per cent of this non-coding DNA is functional, in that it regulates the transcription of other genes. This new way of thinking about 'genes' is important because many DNA associations with complex traits are in these non-coding regions of DNA.

p. 114 *'Each A allele [of the* FTO *SNP] is associated with a three-pound increase in body weight'*: Timothy M. Frayling et al., 'A Common Variant in the FTO Gene is Associated with Body Mass Index and Predisposes to Childhood and Adult Obesity', *Science*, 316 (2007): 889–94. http://doi.org/10.1126/science.1141634.

pp. 114–15 *'This correlation [between the* FTO *SNP and weight] in European populations is 0.09, which accounts for less than 1 per cent of the differences in weight'*: You can square a correlation to find the amount of variance explained. Squaring the correlation of 0.09 between a SNP and a trait indicates that 0.8 per cent of the variance of the trait can be explained by the SNP.

p. 115 *'The possibility of actually correcting a DNA mutation has been realized recently. A gene-editing technique called CRISPR'*: Much has been written about this exciting new technique: Jennifer A. Doudna and Emmanuelle Charpentier, 'The New Frontier of Genome Engineering with CRISPR-Cas9', *Science*, 346 (2014): 1077. doi:10.1126/science.1258096.

p. 117 *'the A allele [of the* FTO *SNP] increases responsiveness to food cues and decreases the extent to which we feel full after eating, or satiety'*: Jane Wardle et al., 'Obesity Associated Genetic Variation in FTO is Associated with Diminished Satiety', *Journal of Clinical Endocrinology and Metabolism*, 93 (2008): 3640–43. doi: 10.1111/j.1469-7610.2008.01891.x.

p. 118 *'There are three steps in the process: getting cells, extracting DNA from the cells and genotyping the DNA'*: The first step, getting cells, can use any cells, because almost all cells have DNA and the DNA is identical in all cells. It's a matter of convenience which cells are obtained. Blood is a good tissue for harvesting lots of DNA, but most often saliva is used because it is easy to collect, even through the post.

The second step is extracting DNA from the cells. Although saliva is more than 99 per cent water, it also contains some sloughed-off cells from our mouths. The cells in our mouth replenish themselves frequently, which is why sores in the mouth heal so quickly. The DNA is physically separated from other stuff in saliva by spinning the saliva in a centrifuge.

The third step is genotyping the DNA. There is not enough DNA for genotyping in the few cells in a saliva sample. For this reason, before genotyping we trick DNA into making millions of copies of itself by hijacking its duplication mechanism.

The process begins by making double-stranded DNA unzip into single strands, which is done simply by heating the DNA. These single strands of DNA are then chopped up into tiny fragments, using enzymes that cut DNA whenever they see a certain DNA sequence.

As happens naturally in the duplication of all cells in our bodies, each single-stranded DNA fragment seeks its complement. In its home environment of the cell there would be lots of A, C, G and T nucleotides floating around, so each single-stranded DNA can form its complement. For SNP genotyping, the DNA fragments are not allowed to combine with individual nucleotides. The fragments are only allowed to combine with short sequences of DNA that we create. The fragments that we create are called probes because they probe for a specific SNP.

Consider the *FTO* SNP on chromosome 16. As mentioned earlier, 15 per cent of us have AA genotypes, 50 per cent AT, and 35 per cent TT. We can probe for this SNP using the non-varying sequence that surrounds the SNP: A-A-T-T-T comes before the A/T SNP and G-T-G-A-T comes after the SNP. We create two single-stranded probes, one with the A allele in the DNA sequence (A-A-T-T-T-**A**-G-T-G-A-T) and the other with the T allele (A-A-T-T-T-**T**-G-T-G-A-T).

Then we turn the single-stranded DNA fragments loose to combine with the single-stranded probes for the A and T alleles. The single-stranded DNA fragments are all tagged with fluorescent labels that light up. Copies of the fragment of chromosome 16 that contain the *FTO* SNP hook up with either the A or T probes. After rinsing away the rest of the DNA that has not found a mate, we can see which probes the DNA fragments combined with. If the DNA fragments fluoresce for the A probe, that means the individual has only the A allele, the A A genotype. If the DNA fragments fluoresce for the T probe, the person has the TT genotype. If the DNA fragments fluoresce for both the A and T probes, this means that the individual's DNA fragments contain both the A and T alleles. Their genotype is AT, indicating that they inherited an A allele from one parent and a T allele from the other parent.

p. 119 *'Many SNPs are very close together on a chromosome and are inherited together as a package'*: That is, they are rarely broken up by recombination, which is a process that occurs during the production of eggs and sperm in which chromosomes exchange parts, described in the Note above.

p. 119 *'This candidate-gene approach did not pay off and led to many false positive findings that did not replicate'*: Christopher F. Chabris et al., 'Most Reported Genetic Associations with General Intelligence are Probably False Positives', *Psychological Science*, 23 (2012): 1314–23. doi: 10.1177/0956797611435528.

Chapter 11: Gene-hunting

p. 121 *'The euphoria of beginning to find genes that predict psychological traits came crashing down as it became clear that none of these reported associations replicated'*: Hundreds of brain-related genes were the focus of thousands of candidate-gene studies of psychological traits during the last three decades. For example, one gene used in many candidate-gene association studies was *COMT* (catechol-O-methyltransferase), which detoxifies stress hormones. A common SNP allele in *COMT* reduces the ability to break down stress hormones in the brain, which results in these hormones floating around for longer. It made sense that this SNP allele might ramp up stress and lead to anxiety and depression. *COMT* was also used as a candidate gene for cognition. In addition to increasing stress in stressful environments, it seemed reasonable to suppose that, in less stressful environments, this SNP allele might improve cognitive function by stimulating the brain.

One problem with the candidate-gene approach is the overly simplistic stories about the function of genes used to justify the selection of a particular gene as a 'candidate'. Every gene does many different things, so it is easy to tell a story about why a gene like *COMT* is a good candidate gene. But these stories are often wrong. Just about any gene could be justified as a candidate for psychological traits because three-quarters of all genes are expressed in the brain.

Another problem is that candidate-gene studies consider only traditional genes, the 2 per cent of the genome that codes for proteins. As indicated earlier, DNA differences that make a difference in psychological traits are usually not in traditional 'genes'. So, candidate-gene studies missed most of the genetic action.

The *COMT* SNP was included in hundreds of studies of cognitive abilities, and even more studies of anxiety. In one of the first candidate-gene studies twenty-five years ago, I set out to compare 100 genes, including *COMT*, in low-IQ versus high-IQ individuals in two independent studies. Although some significant results popped up in the first study, only one replicated in the second study, just what you would expect by chance alone with a *P* value of 0.05. So, the only significant results seemed to be *false positive* findings and I was left empty-handed: Robert Plomin et al., 'Allelic Associations between 100 DNA Markers and High versus Low IQ', *Intelligence*, 21 (1995): 31–48. doi: 10.1016/0160-2896(95)90037-3.

The design I was using had power to detect associations that accounted for more than 2 per cent of the variance of intelligence. Something was wrong here. Perhaps we weren't looking at the right candidate genes. Because we only had power to detect associations that accounted for more than 2 per cent of the variance, another unpalatable possibility was that the effects were smaller than 2 per cent. It turns out the answer was both.

Despite this early warning of negative results for candidate genes, more than 200 subsequent studies reported associations between candidate genes and intelligence. However, most of these involved small samples and there was no attempt to replicate results. In 2012, in a systematic attempt to replicate the top SNPs in twelve candidate genes in three large samples, not a single SNP replicated: Christopher F. Chabris et al., 'Most Reported Genetic Associations with General Intelligence are Probably False Positives', *Psychological Science*, 23 (2012): 1314–23. doi: 10.1177/0956797611435528.

The failure of candidate-gene reports to replicate is not just a problem for research on intelligence. The approach failed everywhere. For example, for schizophrenia, over 1,000 papers reported candidate-gene results for more than 700 genes. A 2015 meta-analysis of the top twenty-four candidate genes found that none replicated: Manillas S. Farrell et al., 'Evaluating Historical Candidate Genes for Schizophrenia', *Molecular Psychiatry*, 20 (2015): 555–62. doi: 10.1038/mp.2015.16.

How can so many published papers have got it so wrong? Earlier, we considered the crisis of confidence in science about failures to replicate. Candidate-gene studies fell prey to all the traps described there. Two of the major pitfalls were that these studies were underpowered and they chased P values.

In relation to the power pitfall, the average sample size of candidate-gene studies was 200. If associations accounted for 5 per cent of the variance, sample sizes of 200 would have adequate power to detect them. But we now know there is not a single effect size anywhere near as large as 5 per cent. The biggest effects are less than 1 per cent. Sample sizes of more than a thousand are needed to detect such small effects.

For this reason, these early candidate-gene studies were at risk of reporting statistically significant results that are not true, or false positives. Scientific journals do not like to publish negative results, so the only results that could be published were reports of positive results, which turned out to be false positives.

The second pitfall was chasing P values, which greatly increases the risk of reporting false positive results. There are several ways that scientists, usually unwittingly, chase P values. They look at several genes or several psychological traits or several ways of analysing the data but only report the results that tell the best story. It is easy to fall prey to this type of cheating because we all want to tell good stories, and this makes it tempting to sweep complications under the carpet. For publication, a good story requires that the results meet the conventional 5 per cent P value. But chasing this P value means that the laws of P (probability) are broken. The chase ends up catching only false positive findings.

There is nothing wrong with trying to tell a good story, as long as the story is true. The problem with the hundreds of candidate-gene stories is that they were not true, yet they led to hundreds of media reports about 'the gene for intelligence' or 'the gene for schizophrenia'. Although candidate-gene studies continue to be published today, most journals now require that papers reporting candidate-gene associations include proof of replication in independent samples prior to publication. False positive findings do not replicate. The hundreds of reports of candidate-gene associations with intelligence and with schizophrenia did not replicate.

The pain of this false start of candidate-gene studies was eased by the success of a new approach that came after the turn of the century, just as it was becoming clear that candidate-gene studies were a flop. The new approach was genome-wide association (GWA), which is the opposite of the candidate-gene approach.

p. 121 *'The dream was to look systematically across the genome rather than picking a few, somewhat arbitrary, candidate genes'*: Neil Risch and Kathleen Merikangas, 'The Future of Genetic Studies of Complex Human Diseases', *Science*, 273 (1996): 1516–17. doi: 10.1126/science. 273.5281.1516. I have not described an older approach to hunting for genes across the genome called linkage analysis. Like genome-wide association, linkage is a systematic genome-wide strategy for gene-hunting. It uses only a few hundred DNA markers across the genome to identify the chromosomal location of major gene effects by examining the co-segregation within family pedigrees between a DNA marker and a disorder. However, linkage is not powerful for detecting smaller gene effects. Linkage can point to the chromosomal neighbourhood, but it cannot pinpoint the exact location. I decided not to discuss linkage, as it is rarely used now because it only has power to detect major-gene effects, whereas most effects are tiny.

p. 121 *'in 1998 I decided to screen the genome, genotyping DNA differences one by one, in order to find DNA differences associated with intelligence'*: Robert Plomin et al., 'A Genome-wide Scan of 1,842 DNA Markers for Allelic Associations with General Cognitive Ability: A Five-stage Design Using DNA Pooling and Extreme Selected Groups', *Behavior Genetics*, 31 (2001): 497–509. doi: 10.1023/A:1013385125887. I reduced the time and money needed by pooling DNA for groups of individuals rather than genotyping each individual separately. This is called *DNA pooling*; it costs no more to genotype 100 individuals than one individual because you pool the DNA for the 100 individuals and genotype the pooled DNA: Lee M. Butcher et al., 'Genotyping Pooled DNA on Microarrays: A Systematic Genome Screen of Thousands of SNPs in Large Samples to Detect QTLs for Complex Traits', *Behavior Genetics*, 34 (2004): 549–55. doi: 10.1023/b%3abege.0000038493.26202.d3.

I compared groups of 100 individuals with high intelligence and 100 individuals of average intelligence. The high-intelligence individuals came from two sources. Half were selected from a larger sample in Cleveland, Ohio, with IQ scores greater than 130. The other half came from a US study that selected adolescents with IQ scores greater than 160. The control sample of individuals with average IQ came from the same Cleveland sample, selecting children with IQs between 90 and 110.

The second shortcut was to use a type of DNA marker with many alleles, because such markers are much more informative than SNPs, which have only two alleles. Simple sequence repeats (SSRs) have many alleles that involve a sequence of two to five base pairs that repeats from five to fifty times, for unknown reasons. The number of repeats is inherited. There are tens of thousands of SSRs in the human genome, mostly in non-coding regions. SSRs are used in DNA fingerprinting, which has revolutionized forensic work by making it possible to create unique DNA profiles for individuals, a DNA 'fingerprint'. We genotyped 2,000 SSRs that are evenly distributed throughout the genome, using a five-stage replication design that weeded out false positive findings. The 2,000 SSRs could not cover every bit of the genome but it could screen a lot of it.

p. 122 *'SNP chips triggered the explosion of genome-wide association studies'*: Joel Hirschhorn and Mark J. Daley, 'Genome-wide Association Studies for Common Diseases and Complex Traits', *Nature Reviews Genetics*, 6 (2005): 95–108. doi: 10.1038/nrg1521.

p. 122 *'the results [of my GWA study on intelligence using SNP chips] were very disappointing'*: Lee M. Butcher et al., 'SNPs, Microarrays and Pooled DNA: Identification of Four Loci Associated with Mild Mental Impairment in a Sample of 6,000 Children', *Human Molecular Genetics*, 14 (2005): 1315–25. doi: 10.1093/hmg/ddi142. We conducted another GWA study, using a new SNP chip with 500,000 SNPs, but found similarly disappointing results: Lee M. Butcher et al., 'Genome-wide Quantitative Trait Locus Association Scan of General Cognitive Ability Using Pooled DNA and 500K SNP (Single Nucleotide Polymorphism) Microarrays', *Genes, Brain and Behavior*, 7 (2008): 435–46. doi: 10.1111/j.1601-183X.2007.00368.x. The top SNP associations from these studies did not replicate: Michelle Luciano et al., 'Testing Replication of a 5-SNP Set for General Cognitive Ability in Six Population Samples', *European Journal of Human Genetics*, 16 (2008): 1388–95. doi: 10.1038/ejhg.2008.100.

p. 122 *'This meant sample sizes not in the hundreds or even thousands but in the tens of thousands'*: The problem was even worse because genome-wide association studies test hundreds of thousands of SNPs throughout the genome. As an extremely conservative correction for multiple testing, it became conventional to correct for 1 million tests in genome-wide association studies. This meant using a P value of not 5 per cent, not 0.5 per cent, but 0.0000005. A sample of 50,000 is needed to have adequate power to detect associations for a quantitative trait like intelligence under these conditions, which seemed impossibly large for psychological research. Worse yet, this is the sample size needed to skim the surface to detect only the very biggest effects. To capture more of the DNA differences responsible for heritability, samples in the hundreds of thousands would be needed.

p. 123 '*In 2007, a GWA study was published that reported analyses of 2,000 cases for each of seven major disorders*': The Wellcome Trust Case Control Consortium, 'Genome-wide Association Study of 14,000 Cases of Seven Common Diseases and 3,000 Shared Controls', *Nature*, 447 (2007): 661–78. doi:10.1038/nature05911.

pp. 123–4 '*the Wellcome Trust Case Control Consortium . . . the only psychological disorder, bipolar disorder, showed no solid SNP associations*': One SNP association was not significant when tested using the usual 'additive' model in which risk increases additively when individuals have one or two risk alleles. The association was only significant when testing a non-additive (recessive) model in which a single risk allele has no effect – the effect only materializes when an individual has two risk alleles. Testing alternative models is reasonable but runs the risk of 'chasing P values', which can, as in this case, run the risk of failing to replicate.

p. 124 '*By 2011 the carping got so bad that ninety-six leading GWA researchers felt it necessary to publish a letter with the title "Don't Give up on GWAS"*': Patrick Sullivan, 'Don't Give Up on GWAS', *Molecular Psychiatry*, 17 (2011): 2–3. doi:10.1038/mp.2011.94.

p. 125 '*Great progress was made during these five years, going from the twenty-four significant associations for seven traits from the Wellcome Trust study to more than 2,000 SNP associations for more than 200 traits*': Peter M. Visscher et al., 'Five Years of GWAS Discovery', *American Journal of Human Genetics*, 90 (2012): 7–24. doi: 10.1016/j.ajhg.2011.11.029.

p. 125 '*After five more years, in 2017, the number of genome-wide significant SNP associations had reached 10,000*': Peter M. Visscher, '10 Years of GWAS Discovery: Biology, Function, and Translation', *American Journal of Human Genetics*, 101 (2017): 5–22. doi: 10.1016/j.ajhg.2017.06.005.

p. 125 '*the Psychiatric Genomics Consortium . . . now includes over 800 researchers from more than 40 countries*': Gerome Breen et al., 'Translating Genome-wide Association Findings into New Therapeutics for Psychiatry', *Nature Neuroscience*, 19 (2016): 1392–6. doi: 10.1038/nn.4411. SNP and twin liability heritabilities are 30 per cent and 80 per cent for schizophrenia, 25 per cent and 90 per cent for bipolar disorder, 20 per cent and 40 per cent for major depressive disorder, 25 per cent and 75 per cent for hyperactivity and 20 per cent and 90 per cent for autism. These SNP liability heritabilities are from: Cross-disorder Group of the Psychiatric Genomics Consortium, 'Genetic Relationship between Five Psychiatric Disorders Estimated from Genome-wide SNPs', *Nature Genetics*, 45 (2013): 984–94. doi: 10.1038/ng.2711. The twin liability heritabilities are from: *Schizophrenia*: Patrick F. Sullivan et al., 'Schizophrenia as a Complex Trait – Evidence from a Meta-analysis of Twin Studies', *Archives of General Psychiatry*, 60 (2003): 1187–92. doi: 10.1001/archpsyc.60.12.1187. *Bipolar disorder*: Nick Craddock and Pamela Sklar, 'Genetics of Bipolar Disorder: Successful Start to a Long Journey', *Trends in Genetics*, 25 (2009): 99–105. doi: 10.1016/j.tig.2008.12.002. *Major depressive disorder*: Patrick F. Sullivan, Michael C. Neale and Kenneth S. Kendler, 'Genetic Epidemiology of Major Depression: Review and Meta-analysis', *American Journal of Psychiatry*, 157 (2000): 1552–62. doi: 10.1176/appi.ajp.157.10.1552. *Hyperactivity*: Stephen V. Faraone and Eric Mick, 'Molecular Genetics of Attention Deficit Hyperactivity Disorder', *Psychiatric Clinics of North America*, 33 (2010): 159–80. doi: 10.1016/j.psc.2009.12.004. *Autism*: Christine M. Freitag, 'The Genetics of Autistic Disorders and Its Clinical Relevance: A Review of the Literature', *Molecular Psychiatry*, 12 (2007): 2–22. doi: 10.1007/s10803-017-3141-1.

p. 125 *'A 2014 [GWA] report from the PGC for schizophrenia included 30,000 cases and netted more than a hundred genome-wide significant associations'*: Schizophrenia Working Group of the Psychiatric Genomics Consortium, 'Biological Insights from 108 Schizophrenia-associated Genetic Loci', *Nature*, 511 (2014) 421–7. doi: 10.1038/nature13595.

p. 125 *'By 2017 the PGC had doubled the number of cases and increased the catch to 155 associations'*: Patrick Sullivan et al., 'Psychiatric Genomics: An Update and an Agenda', *The American Journal of Psychiatry*, 175 (2018) 15–27. doi: 10.1176/appi.ajp.2017.17030283.

p. 125 *'For bipolar disorder . . . [t]he number of genome-wide significant hits has gone from zero to thirty'*: Eli Stahl et al., 'Genome-wide Association Study Identifies 30 Loci Associated with Bipolar Disorder', *bioRxiv* (2017). doi: https://doi.org/10.1101/173062.

p. 125 *'Major depression got off to a slow start, with only one significant hit in a GWA analysis of 20,000 cases'*: Robert A. Power et al., 'Genome-wide Association for Major Depression through Age at Onset Stratification: Major Depressive Disorder Working Group of the

Psychiatric Genomics Consortium', *Biological Psychiatry*, 81 (2017): 325–35. doi: 10.1016/j.biopsych.2016.05.010.

p. 125 *'In 2017 the PGC reported a GWA analysis of over 100,000 cases that identified 44 significant hits'*: Major Depressive Disorder Working Group of the PGC, 'Genome-wide Association Analyses Identify 44 Risk Variants and Refine the Genetic Architecture of Major Depression', *bioRxiv* (2017). doi: 10.1101/167577. In contrast, in 2016, an analysis of 75,000 cases netted 15 significant associations: Craig L. Hyde et al., 'Identification of 15 Genetic Loci Associated with Risk of Major Depression in Individuals of European Descent', *Nature Genetics*, 48 (2016): 1031–6. doi: 10.1038/ng.3623. Another GWA study of 320,000 individuals added individuals who simply reported that they had sought help for depression and found 17 hits: David M. Howard et al., 'Genome-wide Association Study of Depression Phenotypes in UK Biobank (n = 322,580) Identifies the Enrichment of Variants in Excitatory Synaptic Pathways', *bioRxiv* (2017). doi.org/10.1101/168732.

pp. 125–6 *'a recent GWA study of hyperactivity with 20,000 cases reported 12 hits'*: Ditte Demontis et al., 'Discovery of the First Genome-wide Significant Risk Loci for ADHD', *bioRxiv* (2017). doi: https://doi.org/10.1101/145581.

p. 127 *'In 1993 … APOE allele 4 was found to be strongly associated with Alzheimer's disease'*: Elizabeth H. Corder et al., 'Gene Dose of Apolipoprotein E Type 4 Allele and the Risk of Alzheimer's Disease in Late Onset Families', *Science*, 261 (1993), 921–3. doi: 10.1126/science.8346443.

p. 127 *'A 2013 GWA analysis of Alzheimer's disease'*: Jean-Charles Lambert et al., 'Meta-analysis of 74,046 Individuals Identifies 11 New Susceptibility Loci for Alzheimer's Disease', *Nature Genetics*, 45 (2013): 1452–8. doi: 10.1038/ng.2802.

p. 127 *'For psychological disorders, more than a hundred GWA studies have been reported'*: Jacqueline MacArthur et al., 'The New NHGRI-EBI Catalog of Published Genome-wide Association Studies (GWAS Catalog)', *Nucleic Acids Research*, 45 (2017): doi: 10.1093/nar/gkw1133.

p. 128 *'Dimensions provide more power in GWA studies than disorders because every individual counts'*: A GWA study of 50,000 unselected individuals can provide power to detect a SNP association with a trait that accounts for 0.1 per cent of the variance of the trait. For instance, explaining 0.1 per cent of the variance is worth half an IQ point in the familiar intelligence metric of IQ scores, which are standardized to have an average of 100 and a range from 55 to 145 for 99 per cent of the population. But even this tiny effect of 0.1 per cent is not enough.

The next barrier to break will be 0.01 per cent effect sizes (for example, less than .05 of an IQ point), which will require samples of 500,000. Samples of this size are in the pipeline. Reaching this summit of 500,000 individuals, which seems preposterously large for psychological research, will only reveal another, even higher, summit. Samples in the millions will be needed to detect ever smaller effects.

p. 128 *'Another huge advantage of studying dimensions rather than disorders is that the same sample can be used to study many traits'*: For example, most GWA studies of unselected samples include height and weight as anchor variables, which has made it possible to assemble huge sample sizes. Height and weight are archetypes of quantitative traits. Both are highly heritable, 80 per cent for height and 70 per cent for weight. For height, a GWA study of more than 250,000 individuals identified 679 SNPs significantly associated with individual differences in height. For weight, a GWA study of more than 300,000 individuals found 97 hits. The effect sizes of these SNP associations are tiny, with one exception. For weight, one SNP accounted for 1 per cent of the variance, the biggest effect size found for any quantitative trait. This is the SNP in the *FTO* gene described in the previous chapter. The other top SNPs for weight account on average for 0.03 per cent of the differences between people in weight, which translates to effects of 100 grams. Height showed somewhat stronger effects, although the biggest SNP effect was only 0.28 per cent. On average, the top SNPs accounted for 0.07 per cent, which translates to effects of 0.05 cm for height. Andrew W. Wood et al., 'Defining the Role of Common Variation in the Genomic and Biological Architecture of Adult Human Height', *Nature Genetics*, 46 (2014): 1173–86. doi: 10.1038/ng.3097. Adam E. Locke et al., 'Genetic Studies of Body Mass Index Yield New Insights for Obesity Biology', *Nature*, 518 (2015): 197-U401. doi: 10.1038/nature14177.

p. 129 *'The first breakthrough was for an unlikely variable: years of education'*: The first GWA study of years of education was published in 2013: Cornelius A. Rietveld et al., 'GWAS of 126,559 Individuals Identifies Genetic Variants Associated with Educational Attainment', *Science*, 340 (2013): 1467–71. doi: 10.1126/science.1235488. The GWA study was updated in 2016: Aysu Okbay et al., 'Genome-wide Association Study Identifies 74 Loci Associated with Educational Attainment', *Nature*, 533 (2016): 539–42. doi: 10.1038/ng.3552. The next update will include a sample size greater than 1 million, which has identified more than a thousand significant associations: James J. Lee et al., 'Gene Discovery and Polygenic Prediction from a 'Genome-wide Association Study of Educational Attainment in 1.1 Million Individual', *Nature Genetics*, Advance online publication (2018). doi: 10.1038/s41588-018-0147-3.

p. 129 *'Many psychological traits contribute to this heritability, such as previous achievement at school and cognitive abilities, which correlate 0.5 with years of education'*: Eva Krapohl et al., 'The High Heritability of Educational Achievement Reflects Many Genetically Influenced Traits, Not Just Intelligence', *Proceedings of the National Academy of Sciences USA*, 111 (2014): 15273–8. doi: 10.1073/pnas.1408777111.

p. 129 *'For intelligence, GWA studies had only modest success until sample sizes reached almost 300,000, when more than 200 significant associations were reported in 2018'*: The unsuccessful earlier studies with smaller samples have been described: Robert Plomin and Sophie von Stumm, 'From Twins to Genome-wide Polygenic Scores: The New Genetics of Intelligence', *Nature Reviews Genetics*, 19 (2018): 148–159. doi: 10.1038/nrg.2017.104. The most recent GWA study with a sample size of nearly 300,000 is under review: Jennifer E. Savage et al., Genome-wide Association Meta-analysis in 269,867 Individuals Identifies New Genetic and Functional Links to Intelligence', *Nature Genetics*, 50 (2018): 912–19. doi: 10.1038/s41588-018-0152-6.

p. 130 *'For extraversion, a GWA study of 100,000 individuals found 5 hits'*: Min-Tzu Lo et al., 'Genome-wide Analyses for Personality Traits Identify Six Genomic Loci and Show Correlations with Psychiatric Disorders', *Nature Genetics*, 49 (2017): 152–6. doi: 10.1038/ng.3736.

p. 130 *'For neuroticism, over 100 hits were reported in a GWA study with a sample size of 300,000'*: Michelle Luciano et al., '116 Independent Genetic Variants Influence the Neuroticism Personality Trait in over 329,000 UK Biobank Individuals', *bioRxiv* (2017): doi: 10.1101/168906.

p. 130 *For well-being . . . a GWA study of nearly 200,000 individuals found 3 hits*: Aysu Okbay et al., 'Genetic Variants Associated with Subjective Well-being, Depressive Symptoms, and Neuroticism Identified through Genome-wide Analyses', *Nature Genetics*, 48 (2016): 624–32. doi: 10.1038/ng.3552.

p. 130 *'GWA studies of other interesting personality-related traits are popping up'*: Varun Warrier et al., 'Genome-wide Meta-analysis of Cognitive Empathy: Heritability, and Correlates with Sex, Neuropsychiatric Conditions and Brain Anatomy', *bioRxiv* (2017). doi: 10.110 1/081844. Amy E. Taylor and Marcus R. Munafo. 'Associations of Coffee Genetic Risk Scores with Coffee, Tea and Other Beverages in the UK Biobank', *bioRxiv* (2017). doi: 10.1101/096214. Jacqueline

M. Lane et al., 'Genome-wide Association Analyses of Sleep Disturbance Traits Identify New Loci and Highlight Shared Genetics with Neuropsychiatric and Metabolic Traits', *Nature Genetics*, 49 (2016): 274–81. doi: 10.1038/ng.3749. Vincent Deary et al, 'Genetic Contributions to Self-reported Tiredness', *Molecular Psychiatry* (2017). Advance online publication. doi: 10.1038/mp.2017.5. Samuel E. Jones et al., 'Genome-wide Association Analyses in 128,266 Individuals Identifies New Morningness and Sleep Duration Loci', *PLoS Genetics*, 12 (2016): e1006125. doi: 10.1371/journal.pgen.1006125.

p. 131 *'Whole-genome sequencing is the next big thing in genomics'*: Eric D. Green et al., 'The Future of DNA Sequencing', *Nature*, 550 (2017): 179–81. doi: 10.1038/550179a.

p. 131 *'in the next few years a billion individuals will have their whole genome sequenced and this DNA information will be linked to electronic medical records'*: Alkes L. Price et al., 'Progress and Promise in Understanding the Genetic Basis of Common Diseases', *Proceedings of the Royal Society B-Biological Sciences*, 282 (2015): 20151684. doi: 10.1098/rspb.2015.1684.

p. 132 *'most genes affect most brain and behavioural processes'*: A recent paper suggests that genetic effects are not just highly polygenic – they are 'omnigenic' in the sense that most genes will affect most traits: Evan A. Boyle et al., 'An Expanded View of Complex Traits: From Polygenic to Omnigenic', *Cell*, 169 (2017): 1177–86. doi: 10.1016/j.cell.2017.05.038.

p. 133 *'What we know so far is that non-coding regions can be involved in regulation of gene expression'*: It is much more difficult to study gene expression than inherited DNA differences. Gene expression, which begins with the transcription of DNA into RNA, needs to be studied in cells in specific tissues (e.g., brain) at specific ages (e.g., prenatal development) in response to specific environments (e.g., drugs). In contrast, inherited DNA sequence is the same in all cells at all ages in all environments. It is important to remember that all we inherit is DNA sequence. These inherited differences in DNA sequence are responsible for heritability.

p. 133 *'Although the effects of individual SNPs are tiny, these effects can be added like we add items on a test to create a composite score ... I called these SNP sets'*: Lee M. Butcher et al., 'SNPs, Microarrays and Pooled DNA: Identification of Four Loci Associated with Mild Mental Impairment in a Sample of 6,000 Children', *Human Molecular Genetics*, 14 (2005): 1315–25. doi: 10.1093/hmg/ddi142.

Chapter 12: The DNA fortune teller

p. 134 *'After the false start of candidate-gene studies that failed to replicate, GWA research set a stringent criterion for reporting statistically significant "hits" by correcting associations for a million tests across the genome'*: In other words, rather than using the standard P criterion of 5 per cent, correcting for a million tests means that the P value used in GWA studies is 0.00000005.

p. 135 *'SNP genotypes are scored as 0, 1 or 2, indicating the number of "increasing" alleles'*: For example, the *FTO* SNP on chromosome 16 consists of two alleles, T and A. The A allele is associated with a three-pound increase in weight. We each have two alleles for a SNP, one on each of our two chromosomes. Our genotype for the *FTO* SNP can be TT, TA or AA. We can count the number of A alleles in the genotype so that an individual would have a genotypic score of 0, 1 or 2, depending on whether their genotype is TT, TA or AA, respectively. A higher score for this SNP predicts greater body weight. Because each A allele adds three pounds on average, people with the TT genotype are three pounds lighter on average than people with the TA genotype, who are three pounds lighter than people with the AA genotype. This is what is meant by additive genotypic effects – each A allele adds 3 pounds. In addition, like items on any psychological scale, each SNP needs to be added up in the right direction so that the overall polygenic score predicts greater weight. The A allele of the *FTO* SNP happens to be associated with greater weight. For the other SNPs in the GWA analysis of weight, whichever allele is associated with greater weight is counted as 1. Each individual's polygenic score is based on whether the individual has 0, 1 or 2 copies of that allele. Scored in this way, a higher polygenic score predicts greater body weight.

p. 137 *'How many SNPs should go into a polygenic score?'*: In GWA studies, the average effect size of the top SNP associations is about 0.01 per cent. This suggests that polygenic scores need at least 5,000 SNPs to account for heritabilities of 50 per cent if the average effect size is 0.01 per cent. Many more than 5,000 SNPs will actually be required because the effect sizes of the GWA associations include error. Typically, tens of thousands of SNPs are included in polygenic scores. One approach is to keep adding SNPs as long as they increase the power to predict in independent samples: Jack Euesden et al., 'PRSice: Polygenic Risk Score Software', *Bioinformatics*, 31 (2015): 146–8. doi: 10.1093/bioinformatics/btu848. Polygenic scores sometimes include all SNPs: Cornelius A. Rietveld et al., 'GWAS of 126,559 Individuals Identifies Genetic Variants Associated with Educational Attainment', *Science*, 340 (2013): 1467–71. doi: 10.1126/science.1235488. To create my polygenic scores, we used a newer approach, called *LDpred*, which adjusts for the correlation (linkage disequilibrium) between SNPs to avoid 'double counting' correlated SNPs. LDpred also optimizes information from all SNPs, not just the SNPs that are most highly associated with the trait: Bjami J. Vilhjálmsson et al., 'Modeling Linkage Disequilibrium Increases Accuracy of Polygenic Risk Scores', *American Journal of Human Genetics*, 97 (2015): 576–92. doi:10.1016/j.ajhg.2015.09.001.

p. 138 *'To give a sense of the explosion of GWA research during the past decade, the main repository for these results includes GWA summary statistics for 173 traits based on 1.5 million individuals and 1.4 billion SNP-trait associations'*: Zheng et al., 'LD Hub: A Centralized Database and Web Interface to Perform LD Score Regression that Maximizes the Potential of Summary Level GWAS Data for SNP Heritability and Genetic Correlation Analysis', *Bioinformatics*, 33 (2017): 272–9. doi: 10.1093/bioinformatics/btw613.

p. 139 *'In the rest of this chapter, I will share my polygenic scores for height and weight to explore some general issues raised by these indicators'*: My team and I collected my DNA from saliva and extracted the DNA as described earlier. Then we genotyped my DNA for hundreds of thousands of SNPs on a SNP chip. The SNP chip we used was the Illumina Infinium OmniExpress SNP chip, which genotypes 600,000 SNPs across the genome. After quality-control screening, we ended up with 562,199 genotyped SNPs. As is typical, we used these measured SNPs to impute nearby SNPs based on reference panels with whole-genome-sequencing data on large numbers of individuals. Imputation involves inferring SNPs from the reference panels that are highly correlated with (i.e., in linkage disequilibrium with) our measured SNPs. We added 7,323,859 imputed SNPs, which were used together with the measured SNPs to construct my polygenic scores from the results of GWA studies.

After genotyping DNA on a SNP chip, much work is needed to make sense of the raw SNP data. This begins with a series of quality-control analyses that weed out SNP errors. The end product is the creation of hundreds of thousands of SNP genotypes for each individual. These analyses are tedious but are now routine after a decade of work with SNP chips. Not yet routine is the creation of polygenic scores, which have only become widely used in the last two years. The summary statistics for each of the hundreds of thousands of SNPs from a large GWA study for a particular trait are needed to provide the weights to generate polygenic scores for that trait. Many tweaks are being invented to improve polygenic scores, such as taking into account the fact that SNPs close together on a chromosome are correlated.

p. 139 *'The most predictive polygenic score so far is height, which explains 17 per cent of the variance in adult height'*: Derived from summary statistics from: Andrew W. Wood et al., 'Defining the Role of Common Variation in the Genomic and Biological Architecture of Adult Human Height', *Nature Genetics*, 46 (2014): 1173–86. doi: 10.1038/ng.3097. The top SNP associations for height accounted for 0.07 per cent of the variance on average, which translates to effects of 0.05 cm.

p. 141 *'This gap [between the prediction of a trait from a polygenic score and the trait's heritability] is called missing heritability'*: Brendan Maher, 'Personal Genomes: The Case of the Missing Heritability', *Nature*, 456 (2008): 18–21. doi: 10.1038/456018a. Missing heritability is a key issue for all complex traits in the life sciences. Missing heritability is called the 'dark matter' of genome-wide association because, although it certainly exists, we cannot see it. This missing heritability gap will be narrowed as GWA studies become bigger and better. Using current technology, we should be able to more than double the predictive power of polygenic scores with larger GWA samples. Another reason for optimism is that the SNP chips used in GWA studies mostly include common SNPs but most DNA differences are not common. It has been estimated that current SNP chips account for only about half of all the genetic variance in the genome. Teri A. Manolio et al., 'Finding the Missing Heritability of Complex Disease', *Nature*, 461 (2009): 747–53. doi: 10.1038/nature08494. Frank Dudbridge, 'Power and Predictive Accuracy of Polygenic Risk Scores', *PLoS Genetics*, 9 (2013): doi: 10.1371/journal.pgen.1003348.

Because whole-genome sequencing captures all inherited DNA differences, not just common SNPs, it could double the predictive power of polygenic scores. This conclusion is supported by a new method for estimating heritability called *SNP heritability* because it is based on direct DNA measurement of SNPs. SNP heritability estimates the correlation between SNPs and trait similarity for unrelated individuals across the hundreds of thousands of SNPs on a SNP chip. Although there are now several ways to estimate SNP heritability, the first method was called *Genome-wide Complex Trait Analysis (GCTA)*: Jian Yang et al., 'Common SNPs Explain a Large

Proportion of the Heritability for Human Height', *Nature Genetics*, 42 (2010): 565–9. doi: 10.1038/ng.608. Jian Yang et al., 'Genome Partitioning of Genetic Variation for Complex Traits Using Common SNPs', *Nature Genetics*, 43 (2011): 519–25. doi: 10.1038/ng.823.

For complex traits, SNP heritability is generally half the magnitude of twin heritability, which may be due to the fact that current SNP chips only assess common SNPs, whereas most DNA differences in the genome are not common. It has been estimated that current SNP chips tag only about half of the genetic variance: Peter M. Visscher et al., 'Evidence-based Psychiatric Genetics, aka the False Dichotomy between Common and Rare Variant Hypotheses', *Molecular Psychiatry*, 17 (2012): 474–85. doi: 10.1038/m.

There is some evidence that non-SNP DNA differences, rare DNA differences and non-additive genetic effects contribute to missing heritability. In relation to non-SNP DNA differences, copy-number variants have been proposed as a major source of missing heritability: Eric R. Gamazon, Nancy J. Cox and Lea K. Davis, 'Structural Architecture of SNP Effects on Complex Traits', *American Journal of Human Genetics*, 95 (2014): 477–89. doi: 10.1016/j.ajhg.2014.09.009.

In relation to rare variants, rare variants with allele frequencies of less than 5 per cent add 2 per cent to SNP heritability of height: Eirini Marouli et al., 'Rare and Low-frequency Coding Variants Alter Human Adult Height', *Nature*, 542 (2016): 186–190. doi: 10.1038/nature21039. Non-additive genetic variance has also been proposed by some as a major source of missing heritability: Or Zuk et al., 'The Mystery of Missing Heritability: Genetic Interactions Create Phantom Heritability', *Proceedings of the National Academy of Sciences USA*, 109 (2012): 1193–8. doi: 10.1073/pnas.1119675109.

Rare DNA differences have been shown to contribute to risk for schizophrenia, autism and intellectual disability: Fatima Torres, Mafalda Barbosa and Patricia Maciel, 'Recurrent Copy Number Variations as Risk Factors for Neurodevelopmental Disorders: Critical Overview and Analysis of Clinical Implications', *Journal of Medical Genetics*, 53 (2016): 73–90. doi: 10.1136/jmedgenet-2015-103366.

Schizophrenia: David H. Kavanagh et al., 'Schizophrenia Genetics: Emerging Themes for a Complex Disorder', *Molecular Psychiatry*, 20 (2015): 72–6. *Autism*: Michael Ronemus et al., 'The Role of De Novo Mutations in the Genetics of Autism Spectrum Disorders', *Nature Reviews Genetics*, 15 (2014): 133–41. doi: 10.1038/nrg3585. *Intellectual disability*: Lisenka E. L. M. Vissers et al.,

'Genetic Studies in Intellectual Disability and Related Disorders', *Nature Reviews Genetics*, 17 (2016): 9–18. doi: 10.1038/nrg3999. Joep di Light et al., 'Diagnostic Exome Sequencing in Persons with Severe Intellectual Disability', *New England Journal of Medicine*, 367 (2012): 1921–9. doi: 10.1056/NEJMoa1206524.

Another piece of the missing SNP heritability puzzle might be that twin studies overestimate genetic influence: Jian Yang et al., 'Genetic Variance Estimation with Imputed Variants Finds Negligible Missing Heritability for Human Height and Body Mass Index', *Nature Genetics*, 47 (2015): 1114–20. doi: 1038/ng.3390. In addition, more sophisticated statistical methods might be able to narrow the missing SNP heritability gap: Frank Dudbridge, 'Polygenic Epidemiology', *Genetic Epidemiology*, 40 (2016): 268–71. doi: 10.1002/gepi.21966. Huwenbo Shi et al., 'Contrasting the Genetic Architecture of 30 Complex Traits from Summary Association Data', *American Journal of Human Genetics*, 99 (2016): 139–53. doi: 10.1016/j.ajhg.2016.05.013. Douglas Speed et al., 'Re-evaluation of SNP Heritability in Complex Human Traits', *bioRxiv* (2016). doi: 10.1101/074310.

Importantly, SNP heritability, not twin heritability, represents the ceiling for GWA studies, as well as for polygenic scores derived from these GWA studies, because both are limited by the common SNPs assessed on current SNP chips. Robert Plomin et al., 'Common DNA Markers Can Account for More than Half of the Genetic Influence on Cognitive Abilities', *Psychological Science*, 24 (2013): 562–8. doi: 10.1177/0956797612457952.

p. 146 *'Polygenic score for weight'*: Derived from summary statistics from: Adam E. Locke et al., 'Genetic Studies of Body Mass Index Yield New Insights for Obesity Biology', *Nature*, 518 (2015): 197–U401. doi: 10.1038/nature14177. The polygenic score for body mass index (BMI) predicts 6 per cent of the variance. The top SNPS for BMI accounted for 0.03 per cent of the variance on average, which translates to effects of 100 grams.

p. 147 *'polygenic scores for common medical disorders'*: I will mention some of my polygenic scores for medical traits because these traits have had the largest GWA discovery samples. With the GWA data available right now, polygenic profiles can be created for scores of major medical disorders, such as coronary artery disease, Type 2 diabetes, migraine, osteoporosis, rheumatoid arthritis, lung cancer and inflammatory bowel disease. Polygenic scores are also available for many physiological traits, such as cholesterol, triglycerides, insulin sensitivity, resting heart rate, blood pressure and neurological traits.

For many of these disorders, you don't need DNA to find out if you are currently affected. For example, you may know already if you have Type 2 diabetes, high cholesterol or cardiovascular problems. The big difference is that polygenic scores can predict your genetic risk for these disorders, not just assess your current status. If you are overweight and inactive, you are at some risk of Type 2 diabetes. But if you are overweight and inactive *and* have a high genetic risk, your chances are much greater for developing the disorder. What's more, most Type 2 diabetes is not diagnosed until middle age. By then, much of the damage of being overweight and inactive has been done. Knowing your polygenic score earlier in life gives you a better chance to beat the genetic odds by keeping your weight down, eating better and being more active.

Of course, losing weight, eating better and being more active would be good for all of us. But knowing that we are at high risk for Type 2 diabetes is likely to motivate us to actually do it. You can also monitor your blood-sugar levels. Medications can help if diet and exercise are not enough. These are small steps to take and they can't hurt you, at least as compared to doing nothing about your risk for Type 2 diabetes, which can lead to blindness, kidney dialysis and even amputations.

Fortunately, I have only an average polygenic risk for Type 2 diabetes, near the 50th percentile. For Type 2 diabetes, we created my polygenic score based on a GWA study of 25,000 cases that found more than a hundred significant associations: Robert A. Scott et al., 'An Expanded Genome-wide Association Study of Type 2 Diabetes in Europeans', *Diabetes*, 66 (2017): 2888–902. doi: 10.2337/db16-1253.

My polygenic scores for other medical disorders were only somewhat above average. For example, for inflammatory bowel disease, my polygenic score was at the 62nd percentile. For inflammatory bowel disease, we created polygenic scores from a GWA study of 86,000 cases that reported 38 significant associations: Jimmy Z. Liu et al., 'Association Analyses Identify 38 Susceptibility Loci for Inflammatory Bowel Disease and Highlight Shared Genetic Risk across Populations', *Nature Genetics*, 47 (2015): 979–86. doi: 10.1038/ng.3359.

For lung cancer, my polygenic score was at the 67th percentile. For lung cancer, we used a GWA study of 13,500 cases that reported several significant associations: Yesha M. Patel et al., 'Novel Association of Genetic Markers Affecting CYP2A6 Activity and Lung Cancer Risk', *Cancer Research*, 76 (2016): 5768–76. doi: 10.1158/0008-5472.CAN-16-0446.

My polygenic scores were also average for disease-related physiological variables such as resting heart rate (52nd percentile). For resting heart rate, we created polygenic scores from a GWA study of 265,000 individuals that reported 64 significant associations: Ruben N. Eppinga et al., 'Identification of Genomic Loci Associated with Resting Heart Rate and Shared Genetic Predictors with All-cause Mortality', *Nature Genetics*, 48 (2016): 1557–63. doi: 10.1038/ng.3708.

Most of the time, most of us will have scores near the population average. Average scores might seem disappointing, in the sense that they are ambiguous, neither fish nor fowl. However, average scores might be the best outcome. A low polygenic score for a disorder could just mean low risk, which sounds like a good thing. But polygenic scores are always normally distributed, and we don't know what an extremely low score entails. For example, rheumatoid arthritis is an autoimmune disease, which might indicate an overactive immune system, one that sees your own cells as foreign. A very low polygenic score might be a good sign, indicating an immune system less likely to go into overdrive. However, it is also possible that a very low polygenic score indicates other problems. For example, perhaps it indicates a less sensitive immune system that might be more vulnerable to infection.

About rheumatoid arthritis, I was fascinated to learn that my polygenic score for rheumatoid arthritis is at the 96th percentile. Rheumatoid arthritis runs in my family and I am beginning to show some signs of it, especially in my knees. The best preventive action to delay onset is to stop smoking, but I have never smoked. The next best thing is to lose weight, so that's another reason for me to try harder to win my battle of the bulge. Although there is not much I can do about it, I still prefer to know what might be in store for me. If I had known about this risk earlier in life, would I have played less squash, basketball and volleyball, all of which are hard on the knees? If solid scientific evidence told me this made a difference, I probably would have chosen sports nicer on the knees. But there is as yet no such evidence. Now that we can predict genetic risk from early in life, science will have a better shot at finding out how to prevent these problems. Prevention is a much better bet than trying to cure these complex disorders once they occur. My polygenic score for rheumatoid arthritis was based on results from a GWA analysis that included 30,000 cases with rheumatoid arthritis and reported 101 significant associations: Yukinori Okada et al., 'Genetics of Rheumatoid Arthritis Contributes to Biology and Drug Discovery', *Nature*, 506 (2014): 376–81. doi: 10.1038/nature12873.

My polygenic score was also high (87th percentile) for insulin sensitivity, but that's a good thing, because it is thought to be protective against diabetes, although it may also make it more difficult to lose weight. My polygenic score for insulin sensitivity was based on results from a GWA analysis of 17,000 individuals that reported 23 significant associations: Geoffrey A. Walford et al., 'Genome-wide Association Study of the Modified Stumvoll Insulin Sensitivity Index Identifies BCL2 and FAM19A2 as Novel Insulin Sensitivity Loci', *Diabetes*, 65 (2016): 3200–211. doi: 10.2337/db16-0199.

Another interesting medical polygenic score for me was migraine. My polygenic score is at the 83rd percentile. I have had migraines with aura, which are visual symptoms that occur just before the migraine begins. Fortunately, I had them only a couple of times a year as an adolescent and young adult. Now I rarely have them, although I can put myself at risk by staring at my computer screen for too long, with the appearance of aura providing a useful signal that it's time to down tools. We created polygenic scores from a GWA study of 375,000 cases that reported 38 significant associations: Padhraig Gormley et al., 'Meta-analysis of 375,000 Individuals Identifies 38 Susceptibility Loci for Migraine', *Nature Genetics*, 48 (2016): 856–66. doi: 10.1038/ng.3598.

Chapter 13: Predicting who we are

p. 148 *'For schizophrenia, polygenic scores can currently predict 7 per cent of the variance of the liability to be diagnosed as schizophrenic'*: Stephan Ripke et al., 'Biological Insights from 108 Schizophrenia-associated Genetic Loci', *Nature*, 511 (2014), 421–7. doi: 10.1038/nature13595. There is a catch in the phrase 'variance of liability'. GWA analyses of diagnosed disorders rely on comparing individuals diagnosed with the disorder (called cases) versus controls who have not been diagnosed with the disorder. This makes it difficult to talk about variance predicted by the polygenic score, because all that is analysed is the average SNP frequency difference between cases and controls. It is possible to get around this problem statistically by assuming that there is a continuum of liability underlying the dichotomy between cases and controls. The model assumes that individuals are diagnosed as cases when they cross a certain threshold of severity in the continuum of liability. This is called the *liability-threshold model*.

The problem with this model is that one of the 'big findings' of behavioural genetics is that disorders are merely the extremes of the same genetic factors at work throughout the normal distribution. There are no disorders, just dimensions. From this perspective, it seems perverse to assess a dichotomous disorder (cases versus controls) and then assume that it is a continuous dimension.

But the liability-threshold model is reasonable if we think about disorders as the quantitative extremes of normal distributions. Continuing with the extreme example of 'giantism' used earlier, it is as if we took a continuous trait like height and focused on 'diagnosing' giants who are in the top 1 per cent of height. Suppose we did a case-control GWA study of giants versus the rest of the population, throwing away all the information on individual differences in height in the rest of the population. Based on the finding that disorders are merely the extremes of dimensions, results from a GWA study of giants versus controls ought to be similar to those from a GWA study of individual differences in height in the entire population. But why would you compare giants versus the rest of the population when height is so clearly a continuous trait? It doesn't make sense. This is how I think about all disorders – they are merely the quantitative extreme of continuous traits.

For disorders like major depressive disorder, as well as dimensions like height, polygenic scores are perfectly normally distributed as bell-shaped curves. I predict that polygenic scores will hammer more nails into the coffin of diagnostic dichotomies. If the genetic contributions to disorders are normally distributed, it means that, from a genetic perspective, there are no disorders, just dimensions. It is worth being repetitive about this: The genetic differences between people diagnosed with a disorder and the rest of the population are quantitative, not qualitative. There is no threshold where genetic risk tips over into a diagnosable disorder. For continuous dimensions, it is not unreasonable to focus on the extremes, because this is where problems are most severe. But there is no etiologically distinct disorder, just a continuous dimension.

p. 148 *This polygenic score for schizophrenia already predicts more of the liability variance than variables traditionally used to predict risk for schizophrenia*: Evangelos Vassos et al., 'An Examination of Polygenic Score Risk Prediction in Individuals with First-episode Psychosis', *Biological Psychiatry*, 81 (2017): 470–77. doi: 10.1016/j.biopsych.2016.06.028.

p. 149 *'As compared to schizophrenia, current polygenic scores for major depressive disorder and bipolar disorder predict less liability variance – 1 per cent for major depressive disorder and 3 per cent for bipolar disorder'*: Naomi R. Wray et al., 'Genome-wide Association Analyses Identify 44 Risk Variants and Refine the Genetic Architecture of Major Depression', *Nature Genetics*. Advance online publication. doi: 10.1038/s41588-018-0090-3.

p. 150 *'My relatively high polygenic score for schizophrenia makes me even less willing than I would normally be to try the new high-THC forms of cannabis that have been linked to onset of schizophrenia'*: Louise Arseneault et al., 'Cannabis Use in Adolescence and Risk for Adult Psychosis: Longitudinal Prospective Study', *British Medical Journal*, 325 (2002): 1212–13. doi: 10.1136/bmj.325.7374.1212.

p. 151 *'non-diagnosed first-degree relatives of schizophrenics were more likely to be in creative professions, such as actors, musicians and writers'*: Simon Kyaga et al., 'Mental Illness, Suicide and Creativity: 40-Year Prospective Total Population Study', *Journal of Psychiatric Research*, 47 (2013): 83–90. doi: 10.1016/j.jpsychires.2012.09.010.

p. 151 *'people with high polygenic scores for schizophrenia were more likely to be in creative professions'*: Robert A. Power et al., 'Polygenic Risk Scores for Schizophrenia and Bipolar Disorder Predict Creativity', *Nature Neuroscience*, 18 (2015): 953–5. doi: 10.1038/nn.4040.

p. 153 *'The only specific advice would be to avoid head injury – definitely no boxing and probably no heading footballs – because head injury is the one environmental factor known to increase risk for Alzheimer's disease'*: Philip B. Verghese et al., 'Apolipoprotein E in Alzheimer's Disease and Other Neurological Disorders', *Lancet Neurology*, 10 (2011): 241–52. doi: 10.1016/S1474-4422(10)70325-2.

p. 153 *'APOE does most of the heavy lifting for the polygenic score for Alzheimer's disease'*: Valentina Escott-Price et al., 'Polygenic Score Prediction Captures Nearly All Common Genetic Risk for Alzheimer's Disease', *Neurobiology of Aging*, 49 (2017): 214–37. doi: 10.1016/j.neurobiolaging.2016.07.018.

p. 153 *'A polygenic score based on this study predicts more than 10 per cent of the variance in years of education, referred to as educational attainment'*: James J. Lee et al., 'Gene Discovery and Polygenic Prediction from a Genome-wide Association Study of Educational Attainment in 1.1 Million Individuals', *Nature Genetics*, Advance online publication (2018). doi: 10.1038/s41588-018-0147-3.

p. 153 *'A polygenic score based on a GWA study with 330,000 individuals published in 2016 … predicts … 3 per cent of the variance in educational attainment'*: Aysu Okbay et al., 'Genome-wide Association

Study Identifies 74 Loci Associated with Educational Attainment', *Nature*, 533 (2016): 539–42. doi: 10.1038/nature17671.

p. 155 '*A surprising finding from research using the 2016 educational attainment polygenic score is that it predicts intelligence better (4 per cent) than it predicts its GWA target trait of years of education (3 per cent)*': Robert Plomin and Sophie von Stumm, 'From Twins to Genome-wide Polygenic Scores: The New Genetics of Intelligence', *Nature Reviews Genetics*, 19 (2018): 148–59. doi: 10.1038/nrg.2017.104.

p. 155 '*GWA studies of intelligence*': An ongoing GWA analysis of intelligence has reached a sample size of 280,000; its polygenic score predicts 4 per cent of the variance in intelligence: Jeanne E. Savage et al., Genome-wide Association Meta-analysis in 269,867 Individuals Identifies New Genetic and Functional Links to Intelligence', *Nature Genetics*, 50 (2018): 912–19. doi: 10.1038/s41588-018-0152-6. The previous published GWA, with 78,000 individuals, including UK Biobank, yielded a polygenic score that predicts 3 per cent of the variance in TEDS: Suzanne Sniekers et al., 'Genome-wide Association Meta-analysis of 78,308 Individuals Identifies New Loci and Genes Influencing Human Intelligence', *Nature Genetics*, 49 (2017): 1107–12. doi: 10.1038/ng.3869. Earlier GWA studies of intelligence predicted only about 1 per cent of the variance, for example: Gail Davies et al., 'Genetic Contributions to Variation in General Cognitive Function: A Meta-analysis of Genome-wide Association Studies in the CHARGE Consortium (N=53,949)', *Molecular Psychiatry*, 20 (2015): 183–92. doi: 10.1038/mp.2014.188. We conducted a GWA of extremely high intelligence, which yielded a polygenic score that predicts 2 per cent of the variance of intelligence: Delilah Zabaneh et al., 'A Genome-wide Association Study for Extremely High Intelligence', *Molecular Psychiatry* 23 (2018): 1226–32. doi: 10.1038/mp.2017.121.

pp. 155–6 '*We found that the polygenic score created from the results of the 2016 GWA study of total years of schooling in adults predicts 9 per cent of the variance of GCSE scores at the age of sixteen*': Saskia Selzam et al., 'Predicting Educational Achievement from DNA', *Molecular Psychiatry*, 22 (2017): 267–72. doi: 10.1038/mp.2016.107.

p. 156 '*using an approach called* multi-polygenic scores*, we were able to boost this result to predict 11 per cent of the variance in GCSE scores*': A new development in polygenic scores is to combine the predictive power of polygenic scores derived from different GWA studies, called multi-polygenic scores. The rationale behind polygenic scores is to keep adding SNPs from a GWA study until additional SNPs no longer increase the prediction of the target trait in an independent sample. Multi-polygenic scores extend this logic across GWA studies. For example, do the various polygenic scores for intelligence together predict more variance in an independent sample? Even though the relevant GWA studies target different cognitive abilities – reasoning, general intelligence, extremely high intelligence and years of education – their results can be used in a multi-polygenic score analysis. Using this multi-polygenic score approach, we were able to boost the prediction of GCSE scores from 9 per cent to 11 per cent. Eva Krapohl et al., 'Multi-polygenic Score Prediction Approach to Trait Prediction', *Molecular Psychiatry* 23 (2018): 1368–74. doi: 10.1038/mp.2017.203. We also used polygenic scores from the major GWA studies of cognitive-relevant traits in a multi-polygenic score analysis to ask how much variance in intelligence they can predict in TEDS. The polygenic score for years of education by itself predicts 4 per cent of the variance; the other polygenic scores increase this only to 5 per cent. But every little bit counts towards the goal of predicting as much variance as possible: Eva Krapohl et al., 'Multi-polygenic Score Prediction Approach to Trait Prediction', *Molecular Psychiatry* 23 (2018): 1368–74. doi: 10.1038/mp.2017.203. Another study using even more polygenic scores in a multi-phenotypic score predicted 7 per cent of the variance of intelligence in an independent sample: William D. Hill et al., 'A Combined Analysis of Genetically Correlated Traits Identifies 107 Loci Associated with Intelligence', *bioRxiv* (2017). doi: 10.1101/160291. They used a multivariate GWAS approach called *Multi-Trait Analysis of GWAS (MTAG)*: Patrick Turley et al., 'MTAG: Multi-Trait analysis of GWAS', *bioRxiv* (2017). doi: 10.1101/118810.

p. 159 *'Polygenic scores for personality traits were not included because, so far, they do not explain much more than 1 per cent of the variance'*: Aysu Okbay et al., 'Genetic Variants Associated with Subjective Well-being, Depressive Symptoms, and Neuroticism Identified through Genome-wide Analyses', *Nature Genetics*, 48 (2016): 624–32. doi: 10.1038/ng.3552.

Chapter 14: Our future is DNA

p. 163 *'if we had DNA from ourselves as infants and again as adults, the SNP genotypes would be identical and so too would the infant and adult polygenic scores'*: Some random mutations in our DNA occur as time goes by, but the thousands of SNPs that are used to create polygenic scores will not change significantly. DNA can be damaged with aging, especially exacerbated by smoking, but this is also unlikely to affect polygenic scores. Jorge P. Soares et al., 'Aging and DNA Damage in Humans: A Meta-analysis Study', *Aging*, 6 (2014): 432–9. doi: 10.18632/aging.100667.

p. 163 *'When infants are two years old, intelligence tests predict less than 5 per cent of the variance of scores when the individuals are eighteen years old'*: Marjorie Honzik et al., 'The Stability of Mental Test Performance between Two and Eighteen Years', *The Journal of Experimental Education*, 17 (1948): 309–24.

p. 166 *'the effects of large-scale preventive programmes administered in schools, in the community or on the internet are small and temporary'*: Sanne P. A. Rasing et al., 'Depression and Anxiety Prevention Based on Cognitive Behavioral Therapy for at-Risk Adolescents: A Meta-analytic Review', *Frontiers in Psychology*, 8 (2017): Article Number 1066. doi: 10.3389/fpsyg.2017.01066.

p. 166 *'Polygenic scores . . . will inspire a switch of focus to the other, positive, end of the distribution – strengths instead of problems, abilities instead of disabilities, and resiliencies instead of vulnerabilities'*: Robert Plomin et al., 'Common Disorders are Quantitative Traits', *Nature Reviews Genetics*, 10 (2009): 872–8. doi: 10.1038/nrg2670.

p. 168 *'We found in TEDS that the educational attainment polygenic score predicts 5 per cent of the variance in school achievement in secondary school at the age of twelve, and it even predicts 3 per cent of the variance in primary school at the age of seven'*: Saskia Selzam et al., 'Predicting Educational Achievement from DNA', *Molecular Psychiatry*, 22 (2017): 267–72. doi: 10.1038/mp.2016.107.

p. 168 *'a GWA study focused on children's achievement at school could produce polygenic scores that predict several times more variance'*: For example, we found that the EA polygenic score predicted 5 per cent of the variance of reading performance. We showed that a GWA study of reading itself is likely to produce a polygenic score that could explain 20 per cent of the variance of reading performance. Saskia Selzam et al., 'Genome-wide Polygenic Scores Predict Reading Performance throughout the School Years', *Scientific Studies of Reading*, 21 (2017): 334–9. doi: 10.1080/10888438.2017.1299152.

p. 169 *'GWA studies have found genetic correlations greater than 0.5 between schizophrenia, major depressive disorder and bipolar disorder in the PGC, which we replicated in TEDS'*: Cross-disorder Group of the Psychiatric Genomics Consortium, 'Identification of Risk Loci with Shared Effects on Five Major Psychiatric Disorders: A Genome-wide Analysis', *Lancet*, 381 (2013): 1371–9. doi: 10.1016/S0140-6736(12)62129-1. Eva Krapohl et al., 'Phenome-wide Analysis of Genome-wide Polygenic Scores', *Molecular Psychiatry*, 21 (2015): 1188–93. doi: 10.103mp.2015.126.

p. 171 *'the educational attainment polygenic score correlates with parents' socioeconomic status'*: Daniel W. Belsky et al., 'The Genetics of Success: How Single-nucleotide Polymorphisms Associated with Educational Attainment Relate to Lifecourse Development', *Psychological Science*, 27 (2016): 957–72. doi: 10.1177/0956797616643070. Economists and sociologists have become interested in genomics, focusing on socioeconomic outcomes such as income rather than psychological traits. A useful summary of their work can be found in a book by sociologists Dalton Conley and Jason Fletcher, *The Genome Factor* (Princeton University Press, 2017).

p. 171 *'Another twist is that children's own educational attainment polygenic score correlates almost as much with their parents' socioeconomic status. What's more, it also accounts for half of the correlation between family socioeconomic status and children's school achievement'*: Saskia Selzam et al., 'Predicting Educational Achievement from DNA', *Molecular Psychiatry*, 22 (2016): 267–72. doi: 10.1038/mp.2016.107. Eva Krapohl and Robert Plomin, 'Genetic Link between Family Socioeconomic Status and Children's Educational Achievement Estimated from Genome-wide SNPs', *Molecular Psychiatry*, 45 (2015): 2171–9. doi: 10.1038/mp.2015.2.

p. 171 *'the polygenic score for educational attainment explains a significant portion of the correlation between both of these "environmental" measures and children's school achievement'*: Eva Krapohl et al., 'The Nature of Nurture: Multi-polygenic Score Models Explain Variation in Children's Home Environments and Covariation with Educational Achievement', *Proceedings of the National Academy of Sciences, USA*, 114 (2017): 11727–32. doi: 10.1073/pnas.1707178114.

p. 171 *'These are all DNA examples of the nature of nurture, the first studies of this type using polygenic scores'*: Finding genetic influence on environmental measures suggests that GWA studies of environmental measures can yield polygenic scores that predict experience. The first GWA study of an environmental variable was not successful, however, because its sample size was not nearly large enough, given what we now know about how small SNP associations are: Lee M. Butcher and Robert Plomin, 'The Nature of Nurture: A Genome-wide Association Scan for Family Chaos', *Behavior Genetics*, 38 (2008): 361–71. doi: 10.1007/s10519-008-9198-z.

p. 171 *'Genotype–environment interaction is not about the correlation between genes and environments but their interaction'*: Valerie Knopik et al., *Behavioral Genetics*, 7th edition (New York: Worth, 2017).

p. 172 *'The earliest and most famous report of genotype–environment interaction involved an interaction in which a candidate gene's association with antisocial behaviour showed up only for individuals who had suffered severe childhood maltreatment'*: Avshalom Caspi et al., 'Role of Genotype in the Cycle of Violence in Maltreated Children', *Science*, 297 (2002): 851–4. doi: 10.1126/science.1072290.

p. 172 *'Many other interactions between candidate genes and psychological traits have been reported, but most have not replicated'*: Laramie E. Duncan and Matthew C. Keller, 'A Critical Review of the First 10 Years of Candidate Gene-by-environment Interaction Research in Psychiatry', *American Journal of Psychiatry*, 168 (2011): 1041–9. doi: 10.1176/appi.ajp.2011.11020191.

p. 172 *'genotype–environment interaction using polygenic scores'*: In addition to using existing polygenic scores for psychological disorders to investigate whether they interact with treatments, researchers are attempting to conduct GWA studies specifically targeted on interactions between SNPs and treatments, dubbed *GE-Whiz*. That is, rather than looking for SNP associations with the disorder itself, a GE-Whiz GWA analysis looks for SNPs that predict how much individuals respond to the treatment: Duncan C. Thomas et al., 'GE-Whiz! Ratcheting Gene–Environment Studies up to the Whole Genome and the Whole Exposome', *American Journal of Epidemiology*, 175 (2012): 203–7. doi: 10.1093/aje/kwr365. The first GWA study of this type in psychology targeted differences in anxious children's responses to cognitive behavioural therapy, the most effective therapy for anxiety. However, the sample for this pioneering study of 'precision psychology' was too small to yield a reliable polygenic score for genotype–environment interaction: Jonathan R. Coleman et al., 'Genome-wide Association Study of Response to Cognitive-Behavioural Therapy in Children with Anxiety Disorders', *British Journal of Psychiatry*, 209 (2016): 236–43. doi: 10.1192/bjp.bp.115.168229.

p. 172 *'Children in private and grammar schools in the UK have substantially higher educational attainment polygenic scores than students in comprehensive schools'*: Emily Smith-Wooley, 'Differences in Exam

Performance between Pupils Attending Different School Types Mirror the Genetic Differences between Them', *NPJ Science of Learning* (2018). Advance online publication. doi: 10.1038/s41539-018-0019-8.

Another way to look at this is to compare the relative impact on individual differences in school achievement of the EA polygenic score and whether students attend selective or non-selective schools. The EA polygenic score predicts 9 per cent of the variance in GCSE scores, as we have seen. In contrast, school type accounts for 7 per cent of the variance. However, after controlling for selection factors, the variance explained by school type drops to a mere 1 per cent. In other words, the EA polygenic score is nine times more powerful than school type in predicting GCSE scores. In addition, remember that this is the 2016 EA polygenic score, not the upcoming EA polygenic score which should be more than twice as powerful.

p. 173 *'These outcome differences exist, but they are also largely due to pre-existing student characteristics'*: Nida Broughton et al., *Open Access: An Independent Evaluation* (London: The Social Market Foundation, 2014). Available from: http://www.smf.co.uk/wp-content/uploads/2014/07/Open-Access-an-independent-evaluation-Embargoed-00.01-030714.pdf.

p. 174 *'A substantial body of research has shown that educational attainment is heritable'*: Dalton Conley et al., 'Is the Effect of Parental Education on Offspring Biased or Moderated by Genotype?', *Sociological Science*, 2 (2015): 82–105. doi: 10.15195/v2.a6. Benjamin W. Domingue et al., 'Polygenic Influence on Educational Attainment', *AERA Open* 1 (2015): 1–13. doi: 10.1177/2332858415599972.

p. 175 *'Parent–offspring resemblance for educational attainment primarily reflects genetic influence, not environmental inequality'*: Ziada Ayorech et al., 'Genetic Influence on Intergenerational Educational Attainment', *Psychological Science* 28 (2017): 1302–10. doi: https://doi.org/10.1177%2F0956797617707270.

p. 176 *'The Estonian Genome Centre at the University of Tartu created a databank that includes DNA, SNP chip genotypes and extensive data on more than 50,000 Estonians'*: Liis Leitsalu, 'Cohort Profile: Estonian Biobank of the Estonian Genome Center, University of Tartu', *International Journal of Epidemiology*, 44 (2015): 1137–47. doi: 10.1093/ije/dyt268.

p. 177 *'The educational attainment polygenic score predicted twice as much variance of educational attainment and occupational status in the post-Soviet era'*: Kaili Rimfeld et al., 'Genetic Influence on Social

Outcomes during and after the Soviet Era in Estonia', *Nature Human Behaviour* (2018). Advance online publication. doi: 10.1038/s41562-018-0332-5.

p. 177 *'direct-to-consumer companies ... will soon add polygenic score profiles'*: This is beginning to happen: https://DNA.Land.

p. 182 *'Over 80 per cent of those with untreated PKU require twenty-four-hour support and 70 per cent cannot talk beyond single words'*: Glynis H. Murphy et al., 'Adults with Untreated Phenylketonuria: 'Out of Sight, Out of Mind', *British Journal of Psychiatry*, 193 (2008): 501–2. doi: 10.1192/bjp.bp.107.045021.

p. 183 *'the ethical, legal and social implications (ELSI) of the [human genome] project'*: The web page for the National Human Genome Research Institute's ELSI programme: https://www.genome.gov/10001618/. In addition, a recent book covers ethical issues on genomics specifically in relation to education: Susan Bouregy et al. (eds.), *Genetics, Ethics and Education* (Cambridge University Press, 2017).

p. 183 *'I hope that these vexing issues of the costs of personal genomics will be worked out at this level of single-gene medical disorders'*: A helpful discussion of these complex issues can be found in a recent book: Bonnie Rochman, *The Gene Machine: How Genetic Technologies are Changing the Way We Have Kids – and the Kids We Have* (Farrar, Straus and Giroux, 2017).

Acknowledgements

Because *Blueprint* is the culmination of my forty-five-year career, it is tempting to use this opportunity to thank the colleagues, students and friends who have helped shape my career and my research. But there are literally hundreds of them, so I can only hope they know who they are and how much they have meant to me. However, I want to single out for thanks two people whose help was critical in writing this book. Laura Stickney, editor at Penguin, heavily edited the original manuscript and fine-tuned the revised manuscript, in what felt like a master class in transitioning from writing academic papers to writing a book that can reach a wider audience. Sophie von Stumm first suggested that I write this book and encouraged me all along the way, providing the best possible advice and support through many early drafts.

I also want to acknowledge three institutions that have supported my research during the past twenty-five years. Since I came to the UK from the US in 1994, the Medical Research Council has funded my research generously and continuously (MR/M021475/1, previously G0901245) and paid most of my salary as an MRC Research Professor (G19/2). Since 2013, a European Research Council Advanced Investigator Award (295366) has also provided support for my research and part of my salary. Lastly, I am extremely grateful for the opportunity during my twenty-five years in the UK to work in a wonderful interdisciplinary and collegial environment at the Institute for Psychiatry, Psychology and Neuroscience, King's College London.

Index

families – (cont.)
parents see parents/parenting
psychological traits running in
vii, 3
and sibling differences 71–80, 186
socio-economic status 95, 155,
170–71
twin studies see twin studies
and weight see weight: and families
family environment see nurture and
the environment
fatalism 103, 154
Feynman, Richard 33
Fields, W. C. 92
FitzRoy, Robert 79
Fletcher, Jason 217
Franklin, Benjamin 165
Freud, Sigmund vii, 3, 35
friends and peer group
characteristics 44–5
FTO (FaT mass and Obesity-
associated protein) gene 60,
115, 116–17, 128, 132, 136,
137, 192, 220–21, 232

Galton, Francis 79–80
Gattaca 180
generalist genes 66–70, 186
and the brain 69–70
and genetic effects across
cognitive abilities 68–9
and polygenic scores 169
and prediction of psychological
disorders 66–8, 169
genetics
and adoption studies see
adoption studies
behavioural see behavioural
genetics
and biological determinism 53

candidate-gene studies 121, 172,
221–3
as description of what is, not
prediction of what could be 6,
9, 84, 91–2, 103, 154, 193, 195
DNA see DNA
and equal opportunity 93, 94–7,
173–5
and families see families: and
genetics
and fatalism 103, 154
FTO gene see FTO (FaT mass
and Obesity-associated
protein) gene
gene-editing 115–16, 195
gene–environment correlation
96, 98
gene-hunting for DNA
differences 120–33; with
linkage analysis 224
gene expression viii, 113, 115,
119, 133, 231
generalist genes see generalist genes
genetic amplification 57, 207
genetic architecture of
psychopathology 67
genetic clusters 67
genetic influence viii–ix, 3–11,
43; on accidents 47–8; on
children's television viewing
40–44, 171; and climate/
weather 46–7; on divorce
38–40, 89–90; and heritability
see heritability (inherited DNA
differences); as main systematic
force in life xiii, 31, 92, 154,
173, 177; on peer group
characteristics 44–5; on social
mobility 95, 102–5, 173–5; on
social support viii–ix, 45–6,